## PUBLISHER COMMENTARY

Power system planners and decision-makers can use this Guide as they assess potential system vulnerabilities to climate change and extreme weather and develop appropriate resilience solutions. Vulnerability assessments will help utilities determine where and under what conditions their systems may be vulnerable to rising temperatures and sea level, changing precipitation patterns, and more frequent and severe episodes of extreme weather. The resilience plan will identify solutions and prioritize climate resilience actions and investments. By completing the key steps in this Guide, utilities can develop planning-level documents that identify specific actions for managing and mitigating climate change risks.

**Why buy a book you can download for free?  We print this so you don't have to.**

Many documents are distributed only in <u>electronic media</u>.

We at 4th Watch Publishing are former government employees, so we know how government employees actually use the standards.  When a new standard is released, an engineer prints it out, punches holes and puts it in a 3-ring binder.  While this is not a big deal for a 5 or 10-page document, many documents are over 100 pages and printing and binding a large document is a time-consuming effort.  First you have to find a good clean (legible) copy and make sure it's the latest version (not always easy).  Some documents on the web are missing pages or the image quality is so poor, they are difficult to read.  We look over each document carefully and replace poor quality images by going back to the original source document.  We proof each document to make sure it's all there – including all changes.  If you find a good copy, you could print it using a network printer you share with 100 other people (typically its either out of paper or toner).  If it's just a 10-page document, no problem, but if it's 250-pages, you will need to punch 3 holes in all those pages and put it in a 3-ring binder.  Takes at least an hour.

4th Watch Publishing prints these documents so engineers can focus on what they were hired to do – engineering.  This is important because there are not as many engineers working in government as there used to be, so wasted time on clerical duties is unproductive.  It's much more cost-effective to just order the latest version through Amazon.com

**List of Related Publications:**

| | |
|---|---|
| FEMA 577 | Design Guide for Improving Hospital Safety in Earthquakes, Floods, and High Winds |
| FEMA Incident | Incident Management Handbook (IMH) |
| FEMA P-348 | Protecting Building Utility Systems from Flood Damage |
| FEMA | Home Builder's Guide to Construction in Wildfire Zones |
| FEMA | Local Officials Guide for Coastal Construction |
| FEMA | Home Builder's Guide to Coastal Construction |
| FEMA P-55 | FEMA Coastal Construction Manual |
| FEMA P-1000 | Safer, Stronger, Smarter: A Guide to Improving School Natural Hazard Safety |
| FEMA P-453 | Design Guidance for Shelters and Safe Rooms |
| Coast Guard | Coast Guard Incident Management Handbook |
| | National Climate Assessment 2018 |
| Sea Level Rise Maps | Sea Level Rise Maps U.S. East Coast 2100 |
| Sea Level Rise Maps | Sea Level Rise Maps Gulf Coast 2040 – 2100 |
| NIST SP 1205 | Evaluating the Sustainability Performance of Alternative Residential Building Designs |
| NIST TN 1956 | BIRDS v3.1 Low-Energy Residential Database Technical Manual |
| NIST TN 1896 | High-Speed Monitoring of Multiple Grid-Connected Photovoltaic Array Configurations |

# Climate Change and the Electricity Sector:
# Guide for Climate Change Resilience Planning

## September 2016

## U.S. Department of Energy

## Office of Energy Policy and Systems Analysis

## ACKNOWLEDGEMENTS

The U.S. Department of Energy's Office of Energy Policy and Systems Analysis (DOE-EPSA) produced this Guide under the direction of Dr. Craig Zamuda. C.W. Gillespie, Matt Antes, and Paget Donnelly of Energetics Incorporated provided analysis and support in development of the document. Members of the Partnership for Energy Sector Climate Resilience contributed valuable insights and examples based on their work and experiences. The Partnership has 19 member companies: Austin Energy, AVANGRID (formerly Iberdrola USA), Consolidated Edison of New York, Dominion Virginia Power, Entergy Corporation, Exelon Corporation, Great River Energy, Hoosier Energy, National Grid, New York Power Authority, Pepco Holdings Inc., Pacific Gas and Electric, Public Service Electric and Gas, Sacramento Municipal Utility District, San Diego Gas and Electric, Seattle City Light, Southern California Edison, Tennessee Valley Authority, and Xcel Energy. Through this Partnership, DOE-EPSA works with the private sector to develop and deploy effective strategies for enhancing resilience to climate change and extreme weather. The following people reviewed the draft document and provided helpful comments to improve it: Judi Greenwald, Carla Frisch, and James Bradbury, DOE/EPSA; Michael Kuperberg, Fred Lipschultz, and Sarah Zerbonne, USGCRP; and Guenter Conzelmann, Duane Verner, Rao Kotamarthi, and Megan Clifford, Argonne National Laboratory.

## DISCLAIMER

This Guide provides basic assistance to electric utilities and other stakeholders in assessing vulnerabilities to climate change and extreme weather and in identifying an appropriate portfolio of resilience solutions. This document is one component of the U.S. Department of Energy (DOE) response to Executive Order (EO) 13653, *Preparing the United States for the Impacts of Climate Change* (November 2013), which instructs agencies to provide information, data, and tools that local, state, and private-sector leaders can use to improve preparedness and resilience in critical systems—including energy systems. This Guide is also part of a broader DOE effort to inform preparedness, resilience planning, and response initiatives. Related efforts include the following:

- **Partnership for Energy Sector Climate Resilience:** This Partnership consists of 19 utilities, including investor-owned, federal, state, municipal, and cooperative organizations. The goals are to identify best practices, methods, and tools and to accelerate investment in technologies, practices, and policies that will enable a resilient 21st-century energy system. See www.energy.gov/epsa/partnership-energy-sector-climate-resilience.

- **Climate Action Champions:** DOE conducted a national competition to identify local and tribal community organizations pursuing climate change preparedness and resilience activities that can serve as models for other communities. Awardees are working on a range of ambitious activities at the frontier of climate action—from creating climate-smart building codes to installing green infrastructure. See www.energy.gov/epsa/climate-action-champions.

- **State Energy Risk Assessment Initiative:** DOE is collaborating with state and regional organizations to raise state officials' awareness of risk and increase their preparedness to make informed decisions on resilience solutions, energy system and infrastructure investments, energy assurance planning, and asset management. See http://energy.gov/oe/mission/energy-infrastructure-modeling-analysis/state-and-regional-energy-risk-assessment-initiative.

- **State Energy Assurance Plan Assistance:** To increase energy sector resilience, DOE works with state and local governments to develop information and tools and to conduct forums, training sessions, and tabletop exercises for energy officials, emergency managers, policy makers, and industry asset owners and operators. See http://energy.gov/oe/services/energy-assurance/emergency-preparedness/state-and-local-energy-assurance-planning.

In coordination with other federal agencies, DOE is participating in the Climate Data Initiative and contributing to the Climate Resilience Toolkit to provide information, data, and tools that the public and private sectors can use to increase climate change preparedness and resilience. See www.data.gov/climate/energy-infrastructure/.

While these efforts are designed to give electric utilities, regulators, and other stakeholders the information and materials they need to conduct risk-based vulnerability assessments and develop climate change resilience solutions, only a handful of utilities have published climate resilience plans to date. Utilities in the Partnership for Energy Sector Climate Resilience note that managers would welcome additional guidance, tools, and methodologies to help them move forward.

Specific questions may be directed to Craig Zamuda, EPSA, at ClimateResilienceGuide@hq.doe.govtest .

Electric power is essential to nearly all of the critical functions and infrastructures on which modern America relies—from emergency services and communications to transportation, banking, commerce, healthcare, water supply and more. Electricity reliability is increasingly put at risk by climate change and extreme weather events that can exceed the design parameters and other limits of power system assets and operations. Vulnerabilities and feasible solutions vary widely by utility, component, system, region, and geography. Actions taken to improve resilience today, even as a part of routine planning and maintenance, could deliver significant benefits to all users of electricity both now and in the future.

This Guide provides a broad framework for assessing the vulnerability of electric utility assets and operations to climate change and extreme weather and developing appropriate resilience solutions. Vulnerability assessments help utilities to determine where and under what conditions their systems may be vulnerable to rising temperatures and sea levels, changing precipitation patterns, or more frequent and severe episodes of extreme weather. Resilience plans, which are informed by the findings of the vulnerability assessments, identify solutions and prioritize climate resilience actions and investments. By completing the key steps in this Guide (Figure ES.1), utilities will develop planning-level documents that identify specific actions for managing or mitigating climate change risks.

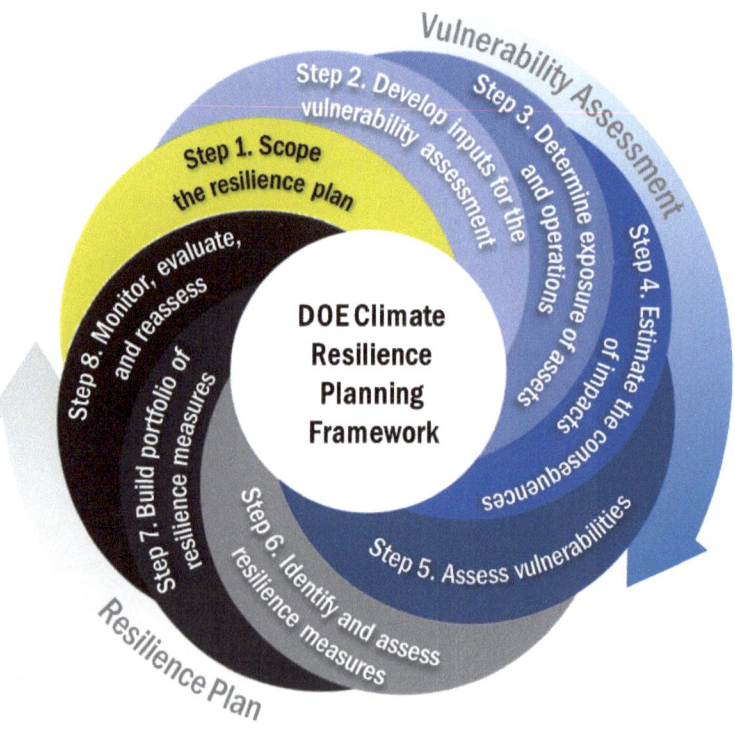

**Figure ES.1.** Steps for conducting a vulnerability assessment and developing climate resilience solutions

## SCOPING THE EFFORT

Step 1 involves defining a useful and practical scope for the climate resilience planning effort. This scope is typically driven by the electric utility's motivations for improving the resilience of its operations and infrastructure. Defining an appropriate scope requires engaging with stakeholders, characterizing the appropriate level of detail for the analysis, identifying key constraints, and determining the types of information and resources that might be needed. Taking the time at the outset to consider all factors driving the resilience planning effort will help utilities hone the scope, as needed, and may facilitate public communication of vulnerability assessment findings and actions outlined in the resilience plan.

## VULNERABILITY ASSESSMENT

Steps 2 and 3 assist utilities in understanding their exposure to climate change and extreme weather hazards. These steps require gathering information on observed trends and future climate projections and taking inventory of potentially vulnerable assets and operations, including supply chains. Using this information and other suggested resources, utilities can more accurately identify relevant hazards and other factors (e.g., geography, region, and hydrology) that may affect the likelihood of potential impacts and the associated severity of any system damages or disruptions.

Step 4 describes methods for calculating the various costs of climate impacts. These costs vary according to the assets or operations affected, the location and severity of the impacts, and the duration of any service disruptions.

Step 5, the final step in the vulnerability assessment, requires a synthesis of the assets and operations exposed to adverse climate events (climate threats), the likelihood and degree of damage or disruption from the climate threats, and the likely consequences if the climate events were to occur (severity of impacts). Exposed assets/operations can be displayed in a likelihood-consequence matrix—a useful visualization tool to help decision makers screen and prioritize risks for the resilience plan.

## RESILIENCE PLAN

The resilience plan relies on information generated or assembled during the vulnerability assessment, such as the likelihood of adverse climate events, the thresholds at which conditions are likely to affect important assets or overall system performance, and the costs or consequences of those adverse climate impacts. The resilience plan prioritizes a set of actions or resilience measures to mitigate critical vulnerabilities. A range of resilience measures may be available either to reduce the probability of damage or disruption (e.g., hardening and relocating assets) or to reduce the business consequences of any damage or disruption (e.g., recoverability and risk transfer/insurance).

Step 6 provides guidance on examining the range of resilience options, determining the costs and impacts of each, and narrowing the selection of actions or measures for inclusion in the plan.

Step 7 assists utilities in determining the most appropriate measures to include in the resilience action plan. This selection process requires a holistic evaluation of the candidate measures, including a comparison of the refined cost/benefit estimates to specified criteria and an assessment of each measure's feasibility, efficacy, co-benefits, and ability to withstand a range or combination of climate impacts.

The resulting resilience plan and associated strategic investment help to ensure that electricity systems will continue to deliver reliable performance in the face of a changing climate. Early action will help utilities maintain their ability to produce and deliver power safely, reliably, and affordably.

## FLEXIBILITY AND IMPROVEMENT

Step 8 provides a framework for monitoring progress, evaluating implementation, and reassessing earlier steps as new information, resources, tools, or technologies become available. Resilience plans must be sufficiently flexible to incorporate new or improved information, including updates on climate change impacts, utility assets, or any other factors affecting system planning and operation.

Recognizing that each electricity system is unique, the Guide sets forth a flexible approach for developing climate resilience plans tailored to the unique needs, goals, and resources of each electric utility and to the mix of climate change impacts and extreme weather events they are likely to encounter. As appropriate at each step, the Guide highlights a range of available tools, projections, sample metrics, and assessments that are now available to assist and guide planners in identifying risks, evaluating options, and developing effective plans.

Building an effective portfolio of resilience measures requires planners to consider both short-term and long-term vulnerabilities and balance tradeoffs. Beyond estimated costs and benefits, resilience plans improve with more detailed or updated information on stakeholder concerns, management objectives, resource availability (natural, human, and financial), science and technology, and other dynamic factors.

Ongoing efforts to address gaps in data, methodologies, tools, and other resources are underway at the U.S. Department of Energy and at academic, government, and industry organizations across the country. Continued communication, data sharing, and coordination on research, best practices, resilience solutions and needs will help leverage resources, strengthen knowledge and projections, and improve resilience.

## TABLE OF CONTENTS

## LIST OF FIGURES

## LIST OF TABLES

# LIST OF ABBREVIATIONS

| | |
|---|---|
| AR5 | Fifth Assessment Report |
| BACEI | Bay Area Council Economic Institute |
| CAIDI | Customer Average Interruption Duration Index |
| CAKE | Climate Adaptation Knowledge Exchange |
| CBA | Cost-Benefit Analysis |
| CDC | Centers for Disease Control and Prevention |
| CDD | Cooling Degree Days |
| CDEEP | Connecticut Department of Energy & Environmental Protection |
| CDF | Customer Damage Function |
| CEC | California Energy Commission |
| CHP | Combined Heat and Power |
| CIAT | Cities Impacts and Adaptation Tool |
| CMI | Customer Minutes of Interruption |
| CMIP | Climate Model Intercomparison Project |
| CMP | Central Main Power |
| ConEd | Consolidated Edison of New York |
| CR-90 | Customer Restoration-90 |
| CSP | Concentrated solar power |
| DCHP | Downscaled CMIP3 and CMIP5 Climate and Hydrology Projections |
| DOE | U.S. Department of Energy |
| DOT | U.S. Department of Transportation |
| EEI | Edison Electric Institute |
| EO | Executive Order |
| EPA | U.S. Environmental Protection Agency |
| EPB | Electric Power Board (of Chattanooga) |
| EPSA | Electric Power Supply Association |
| FEMA | Federal Emergency Management Agency |
| FIRM | Flood Insurance Rate Map |
| FPL | Florida Power & Light Company |
| GAO | Government Accounting Office |
| GCM | General Circulation Model |
| GHG | Greenhouse gas |
| GIS | Geographic Information Systems |
| HDD | Heating Degree Days |
| IAM | Integrated Assessment Model |
| ICE | Interruption Cost Estimation |
| INL | Idaho National Laboratory |
| IPCC | Intergovernmental Panel on Climate Change |
| IRP | Integrated Resource Plan |
| kW | kilowatt |
| kWh | kilowatt-hours |
| kV | kilovolt |
| LBNL | Lawrence Berkeley National Laboratory |
| M&V | Measurement and Verification |

| | |
|---|---|
| MACA | Multivariate Adaptive Constructed Analogs |
| MEA | Maryland Energy Administration |
| MWh | Megawatt-hour |
| NARUC | National Association of Regulatory Utility Commissioners |
| NCA | National Climate Assessment |
| NCCV | National Climate Change Viewer |
| NERC | North American Electric Reliability Corporation |
| NESC | National Electrical Safety Code |
| NGO | Non-governmental Organization |
| NIST | National Institute of Standards and Technology |
| NOAA | National Oceanic and Atmospheric Administration |
| NPCC | New York City Panel on Climate Change |
| NPV | Net Present Value |
| NYPA | New York Power Authority |
| PEPCO | Potomac Electric Power Company |
| PG&E | Pacific Gas & Electric |
| PSC | Public Service Commission |
| RCP | Representative Concentration Pathways |
| RDD&D | Research, Development, Demonstration, and Deployment |
| RDM | Robust Decision Making |
| SAIDI | System Average Interruption Duration Index |
| SAIFI | System Average Interruption Frequency Index |
| SAWTI | Santa Ana Wildfire Threat Index |
| SCE | Southern California Edison |
| SCL | Seattle City Light |
| SDG&E | San Diego Gas & Electric |
| SERDP | Strategic Environmental Research and Development Program |
| SLR | Sea-Level Rise |
| SMUD | Sacramento Municipal Utilities District |
| SRES | Special Report on Emissions Scenarios |
| SRP | Strategic Resource Plan |
| SWCCAR | Assessment of Climate Change in the Southwestern United States |
| TPUC | Texas Public Utilities Commission |
| USFS | United States Forest Service |
| USGCRP | U.S. Global Change Research Program |
| USGS | U.S. Geological Survey |
| VOLL | Value of Lost Load |
| VOS | Value of Service Reliability |
| WCRP | World Climate Research Programme |

# INTRODUCTION

Climate change and extreme weather pose a present and growing threat to the nation's energy systems. In the absence of preventive action, climate change is likely to make our national energy infrastructure increasingly vulnerable to rising temperatures and extreme heat events, wildfires, changing precipitation patterns, more frequent drought, and rising sea levels. The frequency of severe weather events, like intense hurricanes and torrential rains, is also projected to increase. Climate change and extreme weather have the potential to damage energy equipment and facilities, interrupt supply chains and operations, and cause major shifts in energy supply and demand.[1] The resulting disruptions in energy services could adversely affect electric utilities, their customers, and communities, as well as the local and national economy. Across the country, energy systems and infrastructure are already increasingly required to operate outside of the conditions for which they were designed. Appropriate and proactive planning and investment are needed to reduce our energy infrastructure's critical vulnerabilities to climate and extreme weather and to ensure that electric power systems can continue to deliver clean, affordable, and reliable energy with a high level of performance.

## PURPOSE

## PRIMARY USERS

This Guide can assist those involved in making investment decisions, managing risks, ensuring power reliability, administering sustainability plans, or developing infrastructure or operations plans at electric utilities. This document may also be useful to governing bodies that oversee electricity operations and other stakeholders involved in climate change resilience planning. In addition, the information in this Guide could help researchers identify gaps or opportunities in resilience planning tools, methodologies, and technologies, and potentially lead to innovative, cost-effective solutions that enhance climate resilience planning and implementation.

## RATIONALE

Climate hazards are projected to become more frequent and intense in the decades ahead, and extreme weather hazards pose a continuing risk to energy systems. As climate change progresses, energy infrastructures that were built to withstand the known range of historical conditions are becoming more vulnerable to increasingly frequent, intense, and/or sustained heavy precipitation events, extreme temperatures, hurricanes, droughts, wildfires, and rising sea levels.

Resilience planning and strategic investment will help to ensure that electricity systems continue to deliver reliable performance in the face of a changing climate. Early action can help utilities maintain their ability to produce and deliver power safely, reliably, and affordably.

Resilience planning involves assessing the climate vulnerabilities of priority systems and developing an effective action plan to address critical vulnerabilities. Key steps include establishing clear goals; examining the exposure of assets, operations, supply chains, and systems to climate change and extreme weather impacts; and identifying measures that adequately reduce the vulnerability of priority systems or components to these impacts or that reduce the costs of damage or disruption. This process leads to a broader understanding of the climate risks faced by an organization, which, in turn, helps drive informed decision-making and investment in resilience.

## HOW TO USE THIS GUIDE

This Guide presents a systematic, step-by step approach to assessing vulnerabilities and developing a climate resilience action plan. The real-world examples provided suggest the diverse ways in which utilities may collect, process, and act on information at each step. The Guide recognizes that each utility or supplier will have its own set of priorities that must inform the selection of options to improve resilience.

The key analytical steps correspond to each chapter in this guide:

1. Scope the resilience plan
2. Develop inputs for vulnerability assessment
3. Determine exposure of assets and operations
4. Estimate the consequences of climate change impacts
5. Assess vulnerabilities
6. Identify and assess resilience measures
7. Build a portfolio of resilience measures
8. Monitor, evaluate, and reassess the resilience plan

The first step is to establish the scope of the resilience planning effort by identifying the relevant motivations and goals, capabilities and constraints, and stakeholders relevant to the planning process. Steps 2, 3, 4, and 5 of this framework incorporate analytical components that relate to a vulnerability assessment. Steps 6 and 7 address the analytical components of resilience solutions. Finally, Step 8 directs a critical reevaluation of the assumptions and the implementation of prior steps. Under this framework, the vulnerability assessment provides foundational input for the subsequent analysis of actions and investments to increase climate resilience. Some of these actions may have applications or co-benefits beyond enhanced climate resilience—such as improved reliability and reduced greenhouse gas emissions.

The steps outlined in this Guide support a process for continuous improvement. Conducting vulnerability assessments and developing resilience solutions are iterative processes. Information gathered on assets may inform climate information needs, and vice versa. Users should follow the steps in the sequence presented, as each step builds on the previous one. However, as more information becomes available during this process, users may find it useful to repeat entire or individual parts of previous steps.

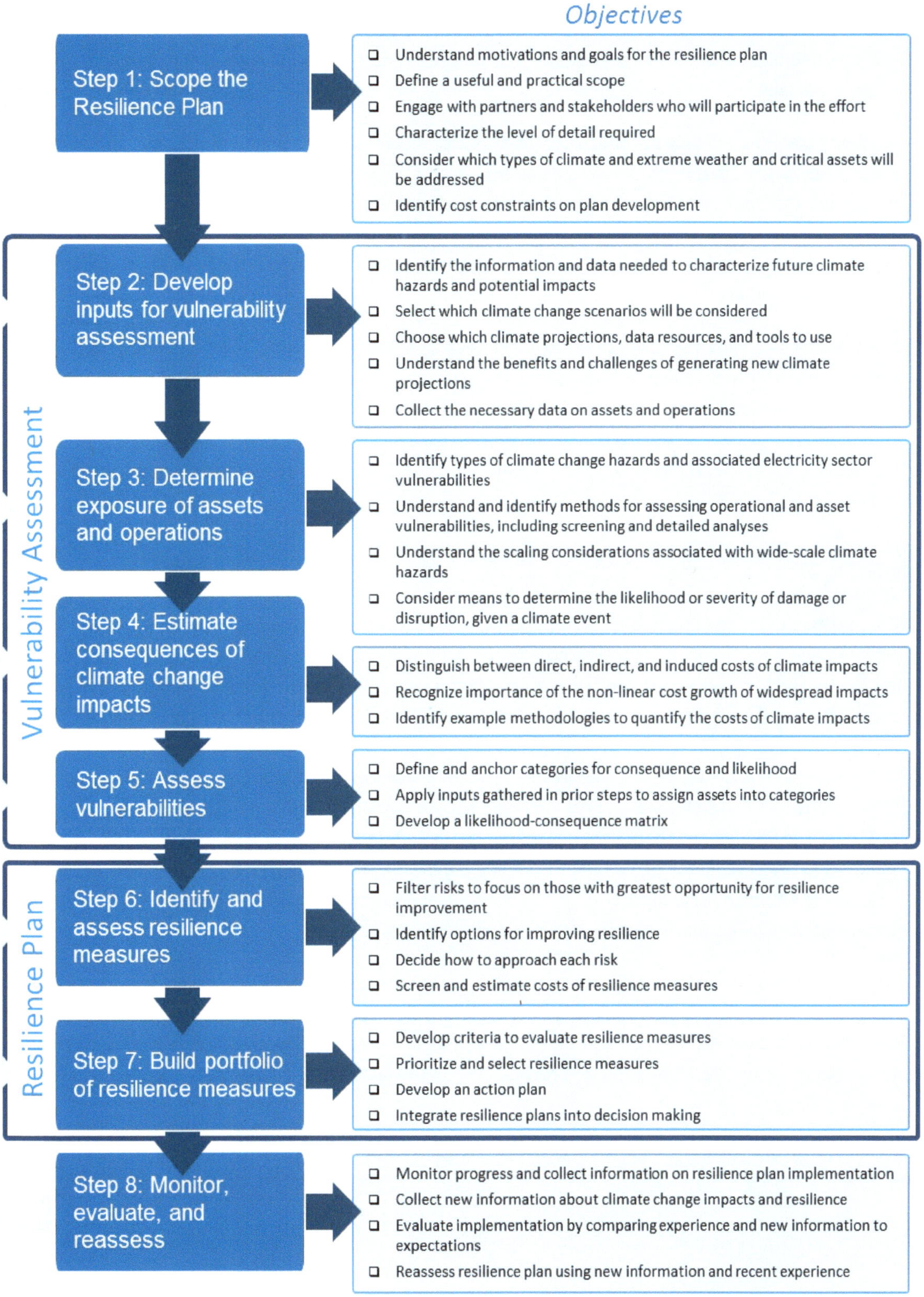

## Objectives

**Step 1: Scope the Resilience Plan**
- ❑ Understand motivations and goals for the resilience plan
- ❑ Define a useful and practical scope
- ❑ Engage with partners and stakeholders who will participate in the effort
- ❑ Characterize the level of detail required
- ❑ Consider which types of climate and extreme weather and critical assets will be addressed
- ❑ Identify cost constraints on plan development

### Vulnerability Assessment

**Step 2: Develop inputs for vulnerability assessment**
- ❑ Identify the information and data needed to characterize future climate hazards and potential impacts
- ❑ Select which climate change scenarios will be considered
- ❑ Choose which climate projections, data resources, and tools to use
- ❑ Understand the benefits and challenges of generating new climate projections
- ❑ Collect the necessary data on assets and operations

**Step 3: Determine exposure of assets and operations**
- ❑ Identify types of climate change hazards and associated electricity sector vulnerabilities
- ❑ Understand and identify methods for assessing operational and asset vulnerabilities, including screening and detailed analyses
- ❑ Understand the scaling considerations associated with wide-scale climate hazards
- ❑ Consider means to determine the likelihood or severity of damage or disruption, given a climate event

**Step 4: Estimate consequences of climate change impacts**
- ❑ Distinguish between direct, indirect, and induced costs of climate impacts
- ❑ Recognize importance of the non-linear cost growth of widespread impacts
- ❑ Identify example methodologies to quantify the costs of climate impacts

**Step 5: Assess vulnerabilities**
- ❑ Define and anchor categories for consequence and likelihood
- ❑ Apply inputs gathered in prior steps to assign assets into categories
- ❑ Develop a likelihood-consequence matrix

### Resilience Plan

**Step 6: Identify and assess resilience measures**
- ❑ Filter risks to focus on those with greatest opportunity for resilience improvement
- ❑ Identify options for improving resilience
- ❑ Decide how to approach each risk
- ❑ Screen and estimate costs of resilience measures

**Step 7: Build portfolio of resilience measures**
- ❑ Develop criteria to evaluate resilience measures
- ❑ Prioritize and select resilience measures
- ❑ Develop an action plan
- ❑ Integrate resilience plans into decision making

**Step 8: Monitor, evaluate, and reassess**
- ❑ Monitor progress and collect information on resilience plan implementation
- ❑ Collect new information about climate change impacts and resilience
- ❑ Evaluate implementation by comparing experience and new information to expectations
- ❑ Reassess resilience plan using new information and recent experience

**Figure 1.** General resilience planning approach for conducting a vulnerability assessment and developing climate resilience solutions.

Some utilities have already developed vulnerability assessments and resilience plans, and the Guide highlights several of those documents as case studies to illustrate the general approach. The document provides descriptions and links to online resources throughout (and further references following each chapter) to help utilities locate a range of available climate change projections and completed vulnerability assessments or utility resilience plans.

Resilience planning will help to reduce potential service interruptions, equipment damage, and associated costs. There is no standardized method for conducting climate resilience planning that will meet all of the needs of all companies. Utilities have a broad range of energy assets, climate- and weather-related risks, and levels of experience with climate change and extreme weather vulnerabilities. Individual assessments and plans will reflect this range and vary widely in terms of detail and analytical depth in characterizing priority vulnerabilities and identifying cost-effective solutions.

Assessments and plans may also range from high-level qualitative assessments for screening purposes to more detailed quantitative and analytical assessments designed to inform asset-specific resilience investment decisions. Users are encouraged to adjust the methodology to support the level of decision-making required by their organization. A successful resilience planning process assumes that users have a solid understanding of their assets and operations, become familiar with the climate stressors in play, and can anticipate how their system may respond.

## KEY SOURCES

The process described in this Guide draws upon existing resources and relevant information developed by DOE or provided by electric utilities, including those that are members of DOE's Partnership for Energy Sector Climate Resilience. The Partnership's regular meetings to discuss and share methodologies, decision tools, and actions for developing and deploying climate-resilient energy technologies have contributed significantly to this document. The Guide also pulls from studies and resources developed by several federal agencies, including the U.S. Department of Energy (DOE), U.S. Department of Transportation (DOT), U.S. Environmental Protection Agency (EPA), and the National Oceanic and Atmospheric Administration (NOAA), among others. Especially useful were DOE's *Climate Change and the Electricity Sector: Guide for Assessing Vulnerabilities and Developing Resilience Solutions to Sea Level Rise*,[2] DOT's *Climate Change & Extreme Weather Vulnerability Assessment Framework*,[3] EPA's *Being Prepared for Climate Change*,[4] and resources from the U.S. Global Change Research Program. This Guide identifies numerous useful resources and directs users to them.

Although utilities have successfully used risk management processes for decades, processes for incorporating climate vulnerabilities and resilience solutions are relatively new. Scientific understanding of climate change projections and potential impacts continues to improve as we learn more about the responses of global and local environments. The DOE-EPSA website[a] provides links to the latest information on energy sector resilience to climate change.

---

[a] http://energy.gov/epsa/office-energy-policy-and-systems-analysis

4

# INTRODUCTION REFERENCES

[1] DOE (U.S. Department of Energy). 2015. *Climate Change and the U.S. Energy Sector: Regional Vulnerabilities and Resilience Solutions*. Washington, DC: DOE. http://energy.gov/sites/prod/files/2015/10/f27/Regional_Climate_Vulnerabilities_and_Resilience_Solutions_0.pdf.

[2] DOE. 2016. *Climate Change and the Electricity Sector: Guide for Assessing Vulnerabilities and Developing Resilience Solutions to Sea Level Rise*. Draft, April.

[3] DOT (U.S. Department of Transportation). 2012. *Climate Change & Extreme Weather Vulnerability Assessment Framework*. Federal Highway Administration. Washington, DC: DOE. December. http://www.fhwa.dot.gov/environment/climate_change/adaptation/publications/vulnerability_assessment_framework/fhwa hep13005.pdf.

[4] EPA (U.S. Environmental Protection Agency). 2014. *Being Prepared for Climate Change: A Workbook for Developing Risk-Based Adaptation Plans*. Office of Water. Washington, DC: DOE. August. https://www.epa.gov/sites/production/files/2014-09/documents/being_prepared_workbook_508.pdf.

## 1. SCOPE THE RESILIENCE PLAN

Informed climate change resilience planning requires a solid understanding of the target infrastructure and operations and their specific vulnerabilities. An important initial step is to identify the primary motivations and goals for conducting the planning exercise. This step will help an electric utility to define a useful and practical scope for the effort, engage with partners and stakeholders, identify cost constraints, characterize the appropriate level of detail for the analysis, and identify the types of data and other information or resources needed to complete the assessment.

### 1.1 IDENTIFY MOTIVATIONS FOR CLIMATE RESILIENCE PLANNING

An electric utility's motivations for climate resilience planning should guide the process from the outset. Taking the time during the early planning stage to consider all of the factors behind the decision to conduct a vulnerabilities assessment and resilience plan

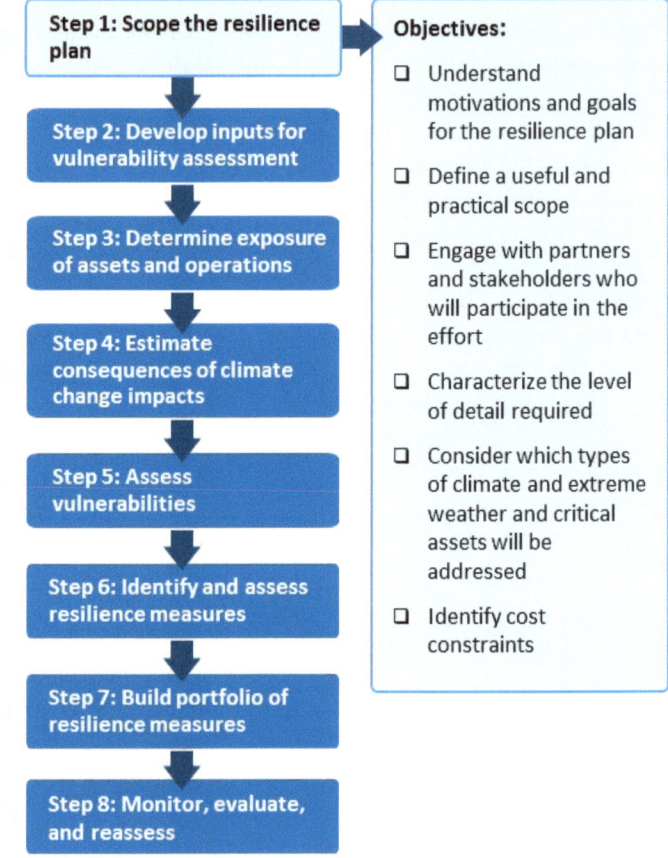

**Step 1: Scope the resilience plan**

**Step 2: Develop inputs for vulnerability assessment**

**Step 3: Determine exposure of assets and operations**

**Step 4: Estimate consequences of climate change impacts**

**Step 5: Assess vulnerabilities**

**Step 6: Identify and assess resilience measures**

**Step 7: Build portfolio of resilience measures**

**Step 8: Monitor, evaluate, and reassess**

**Objectives:**

- ❑ Understand motivations and goals for the resilience plan
- ❑ Define a useful and practical scope
- ❑ Engage with partners and stakeholders who will participate in the effort
- ❑ Characterize the level of detail required
- ❑ Consider which types of climate and extreme weather and critical assets will be addressed
- ❑ Identify cost constraints

will help utilities refine the goals and scope and may facilitate future public communications concerning assessment conclusions. Key motivating questions include the following:

- What past events, incidents, or natural hazards (e.g., storms, outages) may affect decision-making or scope?

- Which stakeholders (e.g., regulators, investors, communities, etc.) are concerned about or interested in climate vulnerabilities? How will the process engage with stakeholders and incorporate input? Which stakeholders will be involved?

- What reports, datasets, tools, or other resources may factor into decision-making? Is this resilience plan driven by the conclusions of specific studies, tools, or other resources?

- Are there any planning gaps that this resilience plan needs to address? How does this assessment fit with other ongoing risk management processes or efforts?

- What types of actions are expected to result from this resilience plan?

- Are any other factors driving the decision to develop a resilience plan?

## 1.2 IDENTIFY RESILIENCE PLAN GOALS

As with motivations, establishing clear goals for the resilience plan will help to define the study's scope, focus the effort, and avoid unnecessary costs or delays. When setting goals, pertinent issues may include the intended use of outputs or conclusions, the nature of the data required (quantitative or qualitative), and any specific questions to be answered. For example, a common goal for a vulnerability assessment is to create a quantifiable estimate of the likelihood and cost of climate impacts for use in a cost-benefit analysis. Identifying goals early in the resilience planning process will allow planners to select the correct tools and methods to provide useful results. Example goals for a resilience plan include the following:

- Identify unknown climate hazards, potential impacts, and associated vulnerabilities.
- Characterize and quantify the probabilities, consequences, and risks associated with known climate vulnerabilities.
- Prioritize vulnerabilities for early response.
- Provide input to evaluations of potential resilience-building actions and measures.
- Provide quantitative inputs to existing risk-management processes.
- Identify risks associated with interconnected utilities, upstream suppliers, and downstream consumers.
- Identify additional stakeholders and increase utility understanding of community goals and concerns.

**Case Study: Consolidated Edison's (ConEd) Motivations and Goals**

Motivated by the widespread damage and costly service disruptions inflicted by Hurricane Sandy, ConEd conducted a 2013 assessment to determine the best way to harden its assets against future storms. Several stakeholders declared the initial report too narrow and claimed the utility had not adequately considered a variety of potential climate change scenarios and hazards, including sea level rise, extreme temperatures, and more intense storms.

As a result of ConEd's rate case before the New York State Public Service Commission (PSC), multiple stakeholders joined with ConEd to create the "Storm Hardening and Resiliency Collaborative" to guide proposed investments in storm hardening. Partners in the Collaborative include representatives of NGOs, such as the Environmental Defense Fund; academia, such as the Columbia Law School Center for Climate Change; New York City's Office of Sustainability; and the New York State Attorney General's Office. At a series of meetings convened by ConEd, members of the Collaborative addressed a range of topics:

- Recommendations on ConEd's storm hardening proposals
- The current design standard for certain ConEd systems—and whether/how to incorporate storm-hardening measures into that standard based on potential climate change impacts
- Development of analytical models to assess the risks and analyze the costs and benefits of proposed storm-hardening projects
- Alternatives to hardening the grid, including resiliency strategies such as microgrids, distributed generation, energy efficiency, and demand response
- Mitigating the climate damages resulting from methane losses from the gas distribution system

Through these discussions, ConEd gained an improved understanding of factors motivating the resilience planning effort and the stakeholders' shared goals. This understanding is now guiding project planning, the creation of a risk assessment and prioritization model, a cost-benefit analysis model, and a climate change study. The NYPSC recently approved the Collaborative's Phase Three report.[1]

## 1.3 DEFINE SCOPE

The scope of a resilience plan should be clearly defined and align with the utility's motivations and goals. Defining scope involves specifying the climate or extreme weather hazards of concern and the types and locations of company assets to be addressed. Scoping also entails determining which systems or components outside of a company's control should be included (typically those critical to the system). Scope further indicates the level of detail appropriate for the assessment, as informed by the utility's goals for the study, available budget and resources, and other factors.

For each step of the resilience planning process outlined in this Guide, utilities may choose to focus the study narrowly (i.e., looking at a subset of assets or hazards, such as electrical sub-stations or coastal flooding), expansively (i.e., examining all assets, operations, and potential impacts related to climate change and extreme weather, including temperature, precipitation, storm/wind, and flooding), or somewhere between those extremes. A narrow scope allows for greater analytical depth and can provide critical insights for certain infrastructures or systems. Such depth can be useful in evaluating a uniquely challenged facility or a system that differs substantially from the rest of a utility's infrastructure (e.g., a single facility or distribution grid in a coastal area). Conversely, a broad scope can help to identify systemic risks and may facilitate a comprehensive approach to integrating climate risk into company-wide risk management practices.

## STAKEHOLDER ENGAGEMENT

Engagement with public stakeholders should be among a utility's highest priorities in resilience planning. If the community does not support the motivations and goals of resilience planning, it may not support the resulting vulnerabilities assessment and resilience investments and actions. The best way to engage stakeholders will vary according to the utility, context, and objectives. One approach is to engage different groups of stakeholders in stages. Successful engagement requires listening, addressing feedback, and offering perspective. The following are strategies for effectively communicating climate change resilience with stakeholders:

- Emphasize that planning for climate change is a best business practice that benefits both customers and utilities.
- Frame resilience as responsible risk management, since preventing impacts is nearly always cheaper than cleaning up and rebuilding after an extreme weather event.
- Explain how the climate affects the geographic area of concern and impacts assets and services that the audience values; use past events, such as a memorable flood or heat wave, to help communicate the meaning of climate change projections.
- Highlight possible solutions to reduce climate risks.

**Case Study: Seattle City Light Stakeholder Engagement**

Seattle City Light (SCL) established a climate initiative to research the impacts of climate change on the utility and develop an adaptation plan to minimize those impacts. Several adaptation activities in SCL's plan involve close engagement with stakeholders. For example, actions include the following:

- Collaborate with resource management agencies and academic institutions to map landslide hazards along SCL's transmission line ROWs, including buffers to accommodate landslides from adjacent land.
- Collaborate with Seattle Public Utilities to evaluate the effects of changes in snowpack and streamflow timing.
- Collaborate with adjacent landowners to reduce use of hazardous fuels and wildfire risk along transmission lines and near critical infrastructure at the hydroelectric projects.
- Collaborate with Skagit Flow Committee, a stakeholder group that can authorize modifications to the flow requirements as necessary to respond to conditions in a given year.[2]

## SELECT ASSETS AND OPERATIONS

A utility may conduct a vulnerability assessment of its entire operation or focus on an individual division or business unit, system, or class of infrastructure. Utilities undertaking a comprehensive vulnerability assessment may benefit from a complete, system-wide understanding of climate hazards, but the scale of such an assessment can be challenging to manage. Conversely, limiting the assessment scope to one or several systems or facilities may achieve greater depth of insight at lower cost but lose the system-wide perspective that might be gained from a comprehensive assessment.

Limiting the scope to enable a more-detailed analysis may make sense for certain assets or operations but not for others, depending upon the type of analysis. For example, a detailed elevation study may be useful for a facility that is sprawled along a sloping shoreline but may be unnecessary for facilities located far from flood threats. In selecting assets and operations for inclusion in the assessment, key considerations include the following:

- Relevance of climate hazards to specific facilities or operations
- Criticality/redundancy of assets or operations
- Relevance of assets/operations to assessment goals
- Expected service lifetime of existing assets or of the planned asset investment

## SELECT CLIMATE HAZARDS AND TIME HORIZON

Identify the climate hazards and potential impacts that the assessment should address and select the timeframe for consideration. A wide range of future changes in climate and extreme weather can significantly affect the energy sector, including several hazards that may not immediately appear relevant (such as small increases in average temperature or seasonal shifts in precipitation patterns that may increase the likelihood of damaging events). Climate-related risks important to a specific electric utility will vary by region and by the asset mix of the utility. Usually, a vulnerability assessment should include all climate change hazards for which reliable projections can be obtained. Planners also need to decide how far into the future projections should be considered as both the timing of projected climate hazards and the expected lifetime of electricity assets vary. For example, examining climate projections over a 50-year horizon may be of limited value for infrastructure nearing retirement but appropriate for guiding investment in a new power plant. The timeframe selected for assessing vulnerabilities should take into account the lifespan of new infrastructure investments. Examples of the kinds of climate-related risks included in power-sector vulnerability assessments are provided in the section below.

## IDENTIFY RELEVANT CLIMATE HAZARDS AND POTENTIAL IMPACTS

Climate hazards include potential events that are driven, enhanced, or affected by the climate and that can damage, destroy, impair, or interrupt energy infrastructure or services. The first step in scoping relevant climate hazards is to review records that may indicate historical exposure to climate and extreme-weather impacts and identify previous events or effects that have caused damages or disruptions. At this scoping stage, the inventory of hazards and potential climate impacts can be expansive and may include hazards that might not ultimately be included in the vulnerability assessment. A review of historical records may include the following:

- Past damages or outages (including distribution, as well as transmission and generation systems, if relevant)
- Spikes in emergency maintenance calls *or* locations with upward-trending maintenance needs
- Price, rate, or demand increases beyond those driven by population and economic factors
- Locations within the system that are affected by or have significant impact on system performance
- Thresholds at which the system begins to experience impacts (e.g., a specific high temperature that has led to elevated probabilities of outages in the past)

With a better understanding of past climate impacts, the next step is to compare these records to prospective climate hazards and the increased risk of impacts that those hazards may impose. Planners should also consider the expected timeframe of projected impacts. Brief descriptions of climate change impacts and their potential to affect the electrical power infrastructure or demand patterns are provided below. Shown another way, Table 1 lists potential climate impacts and implications by energy sector.

> **Increasing Temperature:** In the coming decades, nearly every part of the United States is projected to experience increased temperatures, including both average temperatures and daily highs. Most locations will see a growing number of very hot days and nights and heat waves of increased frequency, length, and severity. Rising temperatures are likely to increase both the average and peak electricity demand for cooling (increasing the number of seasonal cooling degree days; CDDs) and decrease fuel and electricity demand for heating. Higher temperatures can also reduce thermoelectric generation efficiency, transmission and generation capacity, and the service lifetimes of certain equipment (e.g., transformers).

> **Decreasing Water Availability:** Changes in precipitation patterns are projected to vary by region and by season; in some regions, water availability is expected to decrease either annually or during peak demand seasons, raising concerns over water supply. Extreme drought can contribute to increased wildfires, which may cause extensive damage to transmission lines and other energy assets. Changing seasonal precipitation patterns and decreasing snowpack can affect hydropower generation timing and capacity, shifting peak supply from summer into the spring. Similarly, warmer water temperatures and reduced water availability for cooling at thermoelectric facilities could reduce generation capacity. Changes to

precipitation patterns and drought may also affect the price or supply of biomass used to generate electricity.

**Increasing Storms, Flooding, and Sea Level Rise:** Sea levels are already rising along U.S. coastlines, and the rate of sea-level rise is projected to accelerate over the coming century. Rising sea levels pose risks for increased coastal erosion and periodic or permanent inundation of coastal infrastructure. When combined with storm surge, rising sea levels are predicted to increase flooding during coastal storms. Extreme precipitation events are projected to increase in both frequency and intensity in multiple regions of the country. Intense precipitation events pose inland flooding risks for electricity assets and any supporting infrastructure along riverbanks or in floodplains (including other energy systems and critical transportation links between fuel supplies and generation facilities). Atlantic hurricanes, which are expected to increase in intensity over the coming century, are associated with multiple impacts, including intense wind damage, coastal flooding, and wave damage. In addition, extreme winter storm events (e.g., polar vortex and ice storms) can increase physical damage to electricity assets and reduce electricity supply.

**Table 1.** Projected climate change hazards and implications relevant to the energy sector.[3]

| Energy sector | Climate projection | Potential implication |
| --- | --- | --- |
| **Thermoelectric power generation (Coal, natural gas, nuclear, geothermal and solar CSP)** | • Increasing air temperatures | • Reduction in plant efficiencies and available generation capacity |
| | • Increasing water temperatures | • Reduction in plant efficiencies and available generation capacity; increased risk of exceeding thermal discharge limits |
| | • Decreasing water availability | • Reduction in available generation capacity; impacts on coal, natural gas, and nuclear fuel supply chains |
| | • Increasing intensity of storm events, sea level rise, and storm surge | • Increased risk of physical damage and disruption to coastal facilities |
| | • Increasing intensity and frequency of flooding | • Increased risk of physical damage and disruption to inland facilities |
| **Hydropower** | • Increasing temperatures and evaporative losses | • Reduction in available generation capacity and changes in operations |
| | • Changes in precipitation and decreasing snowpack | • Reduction in available generation capacity and changes in operations |
| | • Increasing intensity and frequency of flooding | • Increased risk of physical damage and changes in operations |
| **Bioenergy and biofuel production** | • Increasing air temperatures | • Increased irrigation demand and risk of crop damage from extreme heat events |
| | • Extended growing season | • Increased production |
| | • Decreasing water availability | • Decreased production |
| | • Sea level rise and increasing intensity and frequency of flooding | • Increased risk of crop damage |
| **Wind energy** | • Variations in wind patterns | • Uncertain impacts on resource potential |
| **Solar energy** | • Increasing air temperatures | • Reduction in potential capacity |
| | • Decreasing water availability | • Reduction in concentrating solar potential capacity |
| **Electric grid** | • Increasing air temperatures | • Reduction in transmission efficiency and available transmission capacity |

| Energy sector | Climate projection | Potential implication |
|---|---|---|
| | - More frequent and severe wildfires | - Increased risk of physical damage and decreased transmission capacity |
| | - Increasing intensity of storm events | - Increased risk of physical damage |
| | - Increasing intensity and frequency of flooding | - Disruption of access to remote equipment and facilities |
| Energy demand | - Increasing air temperatures | - Increased electricity demand for cooling; decreased energy demand for heating |
| | - Increasing magnitude and frequency of extreme heat events | - Increased peak electricity demand |
| Fuel transport | - Reduction in river levels | - Disruption of barge transport of crude oil, petroleum products, and coal |
| | - Increasing intensity and frequency of flooding | - Disruption of rail and barge transport of crude oil, petroleum products, and coal |
| | - Thawing permafrost in Arctic Alaska | - Damaged infrastructure and changes to existing operations |
| Oil and gas exploration and production | - Longer sea ice-free season in Arctic Alaska | - Limited use of ice-based infrastructure; longer drilling season; new shipping routes |
| | - Decreasing water availability | - Impacts on drilling, production, and refining |
| | - Increasing intensity of storm events, sea level rise, and storm surge | - Increased risk of physical damage and disruption to offshore and coastal facilities |

When identifying key climate hazards, analysts should consider the availability of reliable projections from existing sources. For example, extreme winds or tornadoes may be important hazards of interest, but current scientific understanding of the relationship between climate change and changes in the frequency or intensity of these hazards is too low to allow actionable projections. In contrast, projections are more readily available with higher degrees of confidence for other trends such as precipitation, temperature increases, and sea level rise.

## IDENTIFY KEY CLIMATE PARAMETERS AND THRESHOLDS

Analysts should identify past infrastructure damages or service disruptions associated with climate impacts and evaluate the extent to which projected changes in the level or duration of various climate parameters may exacerbate future damage or disruptions. To evaluate future risk, utilities should establish evidence-based thresholds at which climate parameters are likely to affect assets or critical resources.

**Case Study: Seattle City Light Uses Climate Thresholds**

Seattle City Light produced a *Climate Change Vulnerability Assessment and Adaptation Plan* that identifies climate parameters for normal operations and thresholds beyond which system operations may become vulnerable. As part of its analysis, SCL determined threshold values for maximum and minimum daily temperatures, CDDs and heating degree days (HDDs), cumulative precipitation, and maximum wind speeds. The utility then evaluated the effects of projected climate changes on each of these parameters. Examples include:

**Precipitation threshold for landslide hazards:** To identify the likelihood of increased landslide hazards, SCL used the U.S. Geological Survey (USGS) cumulative precipitation threshold for issuing landslide warnings: in the Seattle area, landslides are more likely when cumulative precipitation exceeds 3.5 inches over three days or 5.2

inches over 15 days. Using these thresholds, SCL determined that projected increases in both short-term (<24 hour) and total precipitation could increase asset vulnerability to landslides in fall, winter, and spring.

**Wind speed threshold for overhead line damage:** SCL has also applied climate parameter thresholds to boost its adaptive capacity. To prepare for and warn customers about windstorms that could bring down overhead power lines, SCL supported the development of WindWatch. This online tool forecasts high winds in Western Washington up to 72 hours in advance. The tool alerts staff when wind gusts are forecast to exceed the 30 or 40 mph thresholds that signal increased risk of overhead line damage, increasing SCL's capacity to prepare for potentially damaging windstorms. [4]

## SELECT REGION OR GEOGRAPHICAL AREA

Companies operating multiple utilities in different locations or utilities with large service areas may choose to limit the scope of their vulnerability assessment to a particular region. Focusing an assessment on an individual region reduces the complexity of an assessment by limiting the regional variation in climate change projections. Because projected climate threats can vary significantly by region, assessments can be tailored to the geographic area that offers the greatest opportunity to produce actionable information.

### Case Study: Entergy's Focus on Gulf Coast Vulnerabilities

Entergy, recognizing the potential threats to its Gulf Coast assets and operations, worked with several partners to analyze climate vulnerabilities across the region. The results are available in the publication *Building a Resilient Energy Gulf Coast*. Although Entergy's service territory and assets spread across multiple states, including inland states, the assessment focuses exclusively on the 77 counties bordering the Gulf of Mexico. Figure 2 shows the counties examined by the study. This limited geographic scope allowed Entergy and its partners to study the climate hazards unique to the Gulf Region, driven by sea level rise, land subsidence, and increasing hurricane intensity.

The study identified a suite of "no regrets" resilience-building options that offer cost-to-benefit ratios of less than one. It estimates that investing $50 billion in just these options over the next 20 years could avoid $135 billion in losses. An extended suite of resilience measures costing $120 billion could avoid $200 billion in losses over the same period. [5]

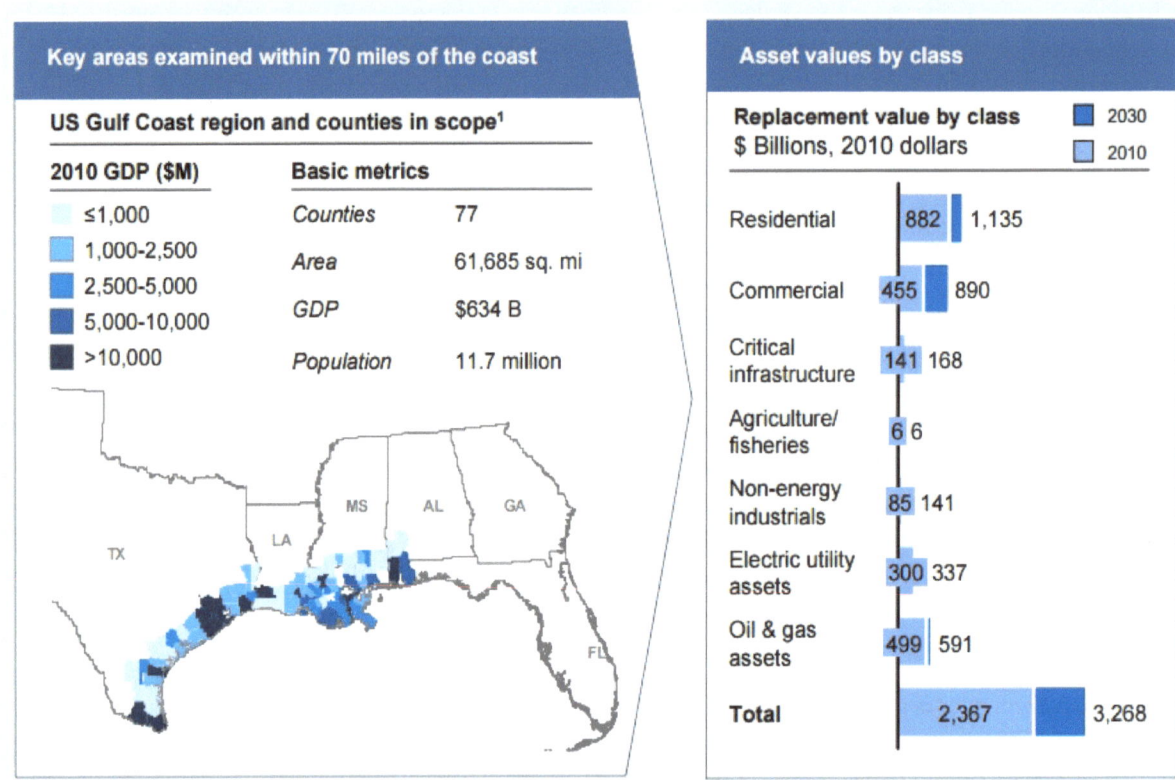

**Figure 2.** Entergy's study, *Building a Resilient Energy Gulf Coast*, limited its scope to those counties near or adjacent to the Gulf Coast and focused on region-specific climate vulnerabilities.[6]

## SELECT EXTERNAL VULNERABILITIES TO ASSESS

A vulnerability assessment need not be limited to the assets and operations of a single utility. In many cases, it may be prudent to investigate the effects of climate change on infrastructure, systems, and sectors that lie outside a utility's fence line or service territory, including critical systems for suppliers and consumers. For example, a utility may want to examine the effects of flooding on any rail, barge or pipeline networks that supply fuel to its power plants.

> **Supplier Vulnerabilities:** Suppliers are any companies that provide products or services to a utility. For utilities with generation assets, suppliers include (but are not limited to) fuel producers and refiners or transportation infrastructure networks such as rail, barge, and pipeline networks. Suppliers may also include generators and transmission systems owned by others.

> **Consumer Vulnerabilities:** For electric utilities, climate changes that affect total or peak consumer demand for electricity may be considered. For utilities with generation or transmission assets, consumers may also include other utilities and institutional consumers.

Utilities can address the vulnerabilities of outside suppliers and consumers by examining the risks of disruption to supply or demand. By stockpiling fuel and identifying backup suppliers, equipment, or mutual aid partners, utilities can prepare for disruptions outside their business operations or service territory. To address consumer vulnerabilities, utilities can maintain close communications and collaborate on climate resilience planning efforts with large industrial and institutional consumers to anticipate and prepare for large changes in demand.

In most cases, the assets and operations of utilities are connected and interdependent, effectively spreading risk over large areas. For example, when a transmission line trips out or a power plant drops offline, connected generation and transmission assets can provide backup capacity and prevent consumer outages. Similarly, mutual aid agreements allow utilities to share restoration crews and reduce the time and cost associated with distribution outages, such as those caused by large storms. Interdependent systems such as mutual aid agreements can increase a utility's resilience to climate impacts by providing response capacity but can also increase a utility's exposure by increasing the likelihood that its crews will be deployed on mutual aid calls. In some cases, interconnections can also expose a utility's territory to systemic risks, such as cascading, large-scale outages—as occurred in the 2003 Northeast Blackout. Anticipating and mitigating systemic risks can be difficult, and may require coordinated state, regional, or national planning efforts.

## IDENTIFY COST CONSTRAINTS ON PLAN DEVELOPMENT

The scope of any vulnerabilities assessment and resilience planning exercise will be constrained by the scale of available funding for developing the plan, including any new tools, resources, research, or any other products produced as a result of the planning process. The costs of plan development are primarily the time and labor of staff and managers tasked with planning, but can also include external consultants and researchers, and licensing costs for data and software tools. In some cases, planning budgets may even include capital, operating, and maintenance costs associated with testing and validating potential resilience measures.

Although it is important to be cognizant of the potential costs of resilience measures prior to beginning the resilience planning process, neither the vulnerabilities assessment nor the resilience plan should be constrained by the expected costs of future resilience measures at the outset of the planning process. The costs associated with projected increases in frequency, intensity and duration of climate hazards during this century challenge both utilities and regulators to identify and approve cost-effective resilience solutions. How regulators address these up-front costs may vary. While in almost all cases, ratepayers will be responsible for covering the costs associated with energy infrastructure upgrades, utilities and regulators are increasingly turning to innovative approaches to deal with funding resilience investments. These include: cost deferral; rate adjustment mechanisms; lost revenue and purchased power adjustments; formula rates; storm reserve accounts; securitization; customer or developer funding/matching contributions; federal funding; and insurance.[7] It can be valuable for both planners and outside stakeholders to understand the constraints that may affect resilience planning early in the process, so that subsequent steps are conducted with them in mind.

# CHAPTER 1 REFERENCES

[1] NYPSC (New York Public Service Commission). 2016. *Order Adopting Storm Hardening and Resiliency Collaborative Phase Three Report Subject to Modifications*, CASE 13-E-0030.
http://documents.dps.ny.gov/public/Common/ViewDoc.aspx?DocRefId={55EA4672-4CA7-409D-A281-0EFD055B083A}.

[2] Seattle City Light. 2016. Seattle City Light Climate Change Vulnerability Assessment and Adaption Plan. Feb 2016.
http://www.seattle.gov/light/enviro/docs/Seattle_City_Light_Climate_Change_Vulnerability_Assessment_and_Adaptation_Plan.pdf.

[3] DOE (U.S. Department of Energy). 2013. *US Energy Sector Vulnerabilities to Climate Change and Extreme Weather*. DOE/PI-0013. Washington DC: DOE. http://energy.gov/sites/prod/files/2013/07/f2/20130716-Energy%20Sector%20Vulnerabilities%20Report.pdf.

[4] Seattle City Light. 2016. Seattle City Light Climate Change Vulnerability Assessment and Adaption Plan. Feb 2016.
http://www.seattle.gov/light/enviro/docs/Seattle_City_Light_Climate_Change_Vulnerability_Assessment_and_Adaptation_Plan.pdf.

[5] Entergy Corporation. 2010. *Building a Resilient Energy Gulf Coast: Executive Report*.
www.entergy.com/content/our_community/environment/GulfCoastAdaptation/Building_a_Resilient_Gulf_Coast.pdf.

[6] Entergy Corporation. 2010. *Building a Resilient Energy Gulf Coast: Executive Report*.
www.entergy.com/content/our_community/environment/GulfCoastAdaptation/Building_a_Resilient_Gulf_Coast.pdf.

[7] Edison Electric Institute. 2014. *Before And After The Storm. A compilation of recent studies, programs, and policies related to storm hardening and resiliency*.
http://www.eei.org/issuesandpolicy/electricreliability/mutualassistance/Documents/BeforeandAftertheStorm.pdf

## 2. DEVELOP INPUTS FOR VULNERABILITY ASSESSMENT

Understanding future climate changes and the exposure of utility assets and operations to climate and extreme weather threats is an information-intensive process. Collecting the information needed to complete a vulnerability assessment helps focus the process and may identify data resources or issues potentially affecting scope. Information requirements include details about utility assets and operations and the existing and projected climate conditions. At this stage, it will also be useful to identify tools, guides, and other resources.

### 2.1 DEVELOP INPUTS ON CLIMATE CHANGE

The assessment will require detailed, localized climate data and information. This should include available data that illustrates observed trends in key climate variables (e.g., extreme precipitation events, heat waves) as well as projections of future climate change for the defined assessment region. Selected

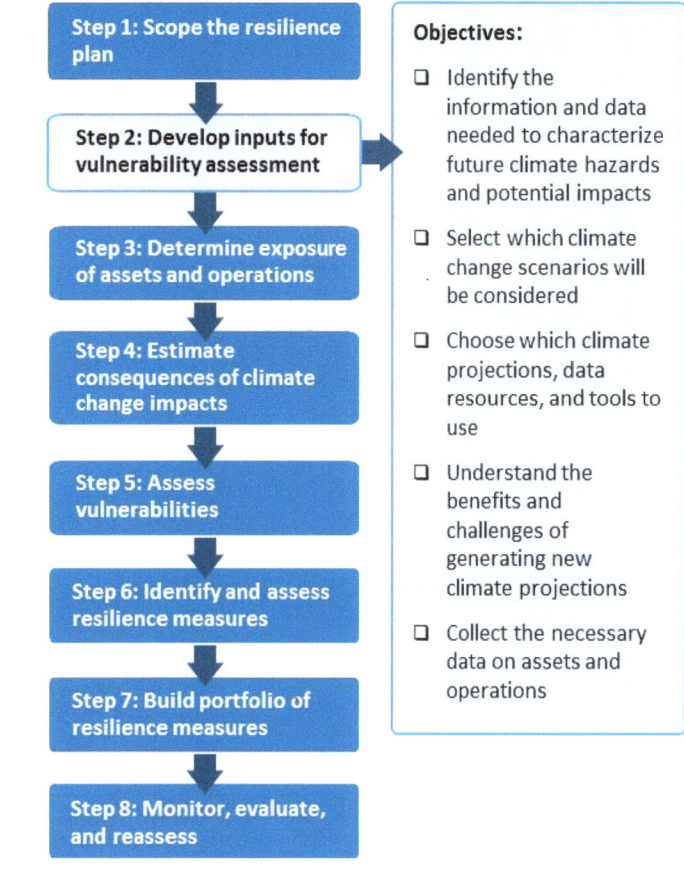

climate projections may include data for multiple different climate scenarios, each based on different assumptions about future emissions levels and the intensity of climate responses. Given the uncertainty about future emissions and the evolving scientific understanding of complex climate processes, it may be prudent to consider multiple scenarios that cover a range of outcomes (e.g., 'high impact,' 'medium impact,' and 'low impact'). The most common scenarios are the Representative Concentration Pathways (RCP) scenarios defined by the Intergovernmental Panel on Climate Change (IPCC) assessment reports. For example, one approach for a vulnerability assessment could be to use the RCP 4.5 as a low emission scenario, and the RCP 8.5 for a high emission scenario. There are many different types of climate change scenarios, and these are discussed in detail in Appendix A.

## IDENTIFY CLIMATE CHANGE PROJECTIONS

A wide range of resources are available to assist utilities in obtaining, processing, and understanding climate change projections. This section provides a summary of a selection of these resources and explains the advantages and disadvantages of the different types of inputs.

## DIRECT CLIMATE MODEL OUTPUTS

Climate change projections originate from multiple sources using different types of methods. The most reliable projections come from coordinated modeling exercises that combine results from different modeling teams, each using a different General Circulation Model (GCMs), to establish a range of possible outcomes. Among these types of exercises, projections from the Coupled Model Intercomparison Project (CMIP) conducted by the World Climate Research Programme (WCRP) are well regarded. The CMIP projections are used as inputs for the IPCC's Assessment Reports, and the latest CMIP projections are called CMIP5. Sources for CMIP projections include the following:

**Downscaled CMIP3 and CMIP5 Climate and Hydrology Projections (DCHP):** The DCHP website hosts a large collection of WCRP climate change projections that have been downscaled for the contiguous United States. The DCHP project is a collective effort by several U.S. federal agencies, institutions, and organizations. DCHP projections include individual model runs for the CMIP3 and CMIP5 scenarios and models, and each is downscaled using several different methods. The DCHP website provides additional context and some tutorials on how to use the data — http://gdo-dcp.ucllnl.org/downscaled_cmip_projections/dcpInterface.html[1]

**MACA Downscaled CMIP5 Projections:** As another source of downscaled CMIP5 projections, the University of Idaho hosts a selection of CMIP5 model runs downscaled using a different method (called Multivariate Adaptive Constructed Analogs, or MACA) — http://maca.northwestknowledge.net/index.php

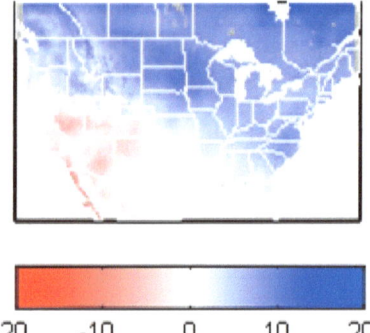

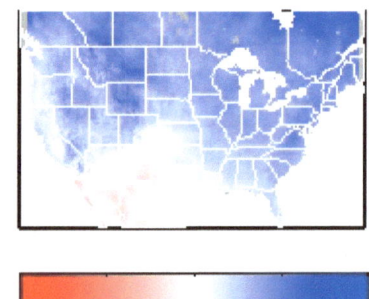

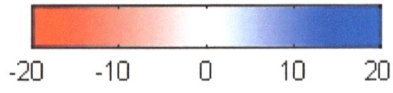

**Figure 3.** DCHP downscaled precipitation projections demonstrate the difference between CMIP3 and CMIP5.[1]

**CMIP5 SERDP Downscaled Projections:** The Department of Defense, EPA, and DOE collaborate on the Strategic Environmental Research and Development Program (SERDP), which has produced dynamically downscaled projections using three different climate models from CMIP5 and the WRF regional-scale model. Future time slices for mid-century 2045-2055 and end of the century 2085-2095 are also available for three climate models and two different forcing scenarios, RCP8.5 and RCP 4.5. The dataset is scheduled to be made available on a portal at Argonne National Laboratory.

Additional climate change projections and other climate-related data can be found via the Climate Data Initiative at https://www.data.gov/climate/.

It is important to recognize that while climate models can provide valuable insights into future climate trends that could affect a particular region, there is never a perfect match between model simulations and observed climate conditions. Because climate models simulate atmospheric and natural processes at the global scale, systemic biases may arise at the regional, or local-scale. For this reason, it is necessary to be aware of the potential for local biases when using downscaled climate projections, and to identify and correct for any biases before using the projections for vulnerability assessment and planning purposes.[2] Utilities seeking to use direct climate model

18

outputs should review the technical guidance associated with the downscaled projections. One useful report, *Use of Climate Information for Decision-Making and Impacts Research: State of Our Understanding* is especially suited to this purpose.[3]

## BOUNDED VS. PROBABILISTIC CLIMATE PARAMETERS

Steps 2 through 5 of this Guide provide a framework for completing a vulnerabilities assessment using multiple climate scenarios to establish a range of potential outcomes for climate parameters such as average temperature, precipitation, and sea level rise. This "bounded parameters" approach is an effective and efficient means of identifying the potential exposure of assets and operations to climate hazards, considering the additional effort required with more exhaustive alternative approaches and the associated uncertainty in detailed future projections. However, a more advanced assessment that estimates probability distributions for each climate parameter and scenario could improve risk-based decision-making by allowing probabilistic estimates of the likelihood of specific climate outcomes. With currently available tools, this type of assessment would require custom modeling or downscaling of climate parameters.[b]

Although more complex and labor-intensive, one advantage of a probabilistic treatment of climate parameters is that when combined with historical climate and extreme weather trends, these data can be used to create functional-form risk estimates that utilize both the distribution of climate parameters and attached costs. Assessments that use two or three climate scenarios without probabilistic climate parameters may not effectively represent the full range of potential outcomes, since the upper and lower bounds are the mean of the distributions of outcomes in each scenario. Bounded risk estimates cannot provide quantitative probabilities of climate outcomes, only qualitative bounds such as "high" and "low" bounds. Moreover, probabilistic risk assessments can buttress cost-benefit analysis (Chapter 7).

## ASSESSMENT PRODUCTS AND ANALYSIS TOOLS

Utilities may be able to take advantage of existing assessment reports or other products that provide pertinent climate projections along with high-level analysis of the projections. These published products synthesize the results of multiple scientific papers to provide a coherent message about possible future climate changes. Assessment products can be based on a custom set of climate simulations for the national, regional, state, or local scale. If such an integrated climate analysis has been completed and scaled for an area or region close to the assessment boundaries, it is likely to serve as a valuable resource for the assessment.

The advantage of relying on assessment products is that sizeable effort has already gone into ensuring that the climate model results are presented accurately and within the context of other projected changes, account for and correct any model biases, and that they reflect the appropriate levels of uncertainty about both model capabilities and natural processes. When using raw climate model outputs, it can be easy to misinterpret individual projections or miss important context. For example, the potential drying effects of projected increases in temperature may outweigh projected future increases in precipitation. For this reason, planners who are unsure how to approach

---

[b] Ongoing efforts by the U.S. Department of Energy and others aim to provide downscaled, county-level climate projections with probability density functions defined for each climate parameter.

climate projections may use those provided in the U.S. National Climate Assessments or the IPCC Assessment Reports.

A key disadvantage to using existing assessment products is that one may not be available for a utility's specific region of focus. If national or global-scale climate change projections are the only available climate data, detailed local assessment of asset vulnerabilities is likely to be more difficult. Moreover, some state or regional-level assessment products may not consider the full range of climate scenarios or model types.

A variety of example assessment resources are briefly described below, including national, regional, and local reports:

## NATIONAL AND GLOBAL ASSESSMENT RESOURCES

**USGCRP Third *National Climate Assessment* (NCA):** The United States Global Change Research Program (USGCRP) is a subprogram of the White House National Science and Technology Council charged with providing actionable assessments of climate change science for the United States. The USGCRP's signature product, the NCA, is now in its third edition. The third NCA summarizes climate change hazards for 10 regions across the United States as well as for 13 sectors or ecosystems. It is one of the most comprehensive and useful reports on potential climate change impacts in the United States. See http://nca2014.globalchange.gov/report.

**NOAA Regional Climate Trends and Scenarios for the U.S. National Climate Assessment:** NOAA developed the Regional Climate Trends series as a set of inputs for the third NCA, but it is also useful as a stand-alone review of observed climate trends and set of detailed climate projections for the contiguous United States. Because these projections are not part of an assessment report, they should be used alongside the NCA to gain context for the projections. When used together, these resources can provide locally detailed climate projections as well as useful deployment information. See www.nesdis.noaa.gov/technical_reports/142_Climate_Scenarios.html.

**IPCC Fifth Assessment Report (AR5):** The IPCC's AR5 is the premier assessment resource for global climate change projections. The assessment report synthesizes scientific literature on climate science and impacts and provides the best scientific estimate of climate change impact probabilities. Due to its global scope, the AR5 does not provide projections with high geospatial resolution. The AR5 report is divided into three sections (called working groups): the Physical Science Basis; Impacts, Adaptation and Vulnerability; and Mitigation of Climate Change. See www.ipcc.ch/report/ar5/.

## LOCAL AND REGIONAL ASSESSMENT RESOURCES

**Austin, TX:** The City of Austin commissioned a report on climate change projections to facilitate vulnerability assessments for city services. These projections were completed in 2014 and are available along with the city's resilience planning materials on the website of the Austin Office of Sustainability at http://austintexas.gov/page/climate-resilience.

**California:** The California Climate Change Center's latest assessment report was released in 2012. It includes statewide impact projections, including a focus on energy systems and a regional focus on the

San Francisco Bay Area. Visit
www.climatechange.ca.gov/climate_action_team/reports/third_assessment/index.html.

**New York City, NY:** The New York City Panel on Climate Change (NPCC) has produced a series of reports on climate hazards in the region. NPCC's latest report, *Building the Knowledge Base for Climate Resiliency,* was released in 2015. See www.nyas.org/Publications/Annals/Detail.aspx?cid=5c5c2bdd-795f-4904-acd5-e3fe4a5c338a.

**Philadelphia, PA:** To inform resiliency planning, Philadelphia commissioned the 2014 report *Useful Climate Information for Philadelphia: Past and Future.* See www.phila.gov/green/pdfs/UsefulClimateScience.pdf.

**Southwest:** The Assessment of Climate Change in the Southwestern United States (SWCCAR) is a collaborative regional climate assessment report completed by the Southwest Climate Alliance in 2013. It includes regionally relevant climate projections and details on potential impacts for six southwestern states: Arizona, California, Colorado, Nevada, New Mexico, and Utah. The complete report is available online at www.swcarr.arizona.edu.

**Washington:** The Washington State Climate Resources Clearinghouse includes links to relevant assessment materials in the 2012 report *Washington State's Integrated Climate Response Strategy* and the 2009 *Comprehensive Assessment of Climate Change Impacts on Washington State.* See www.ecy.wa.gov/climatechange/ipa_resources.htm.


**Case Study: Sacramento Municipal Utilities District (SMUD) identifies key climate risks from California Climate Change Assessment**

In its 2012 report, *Climate Readiness Strategy,* SMUD looked to strengthen and update its understanding of the likely impacts of climate change on utility systems in the Sacramento area. SMUD drew on the best local research to assess how its operations and facilities may be affected by future changes in key climate-related parameters, including regional hydrology, wind, and wildfire. The Second California Climate Change Assessment, which served as a key input for the study, identifies significant potential vulnerabilities and key areas of uncertainties for future analysis. SMUD's report draws from historical records of relevant weather parameters and looks at projected changes, as derived from existing scientific literature and various climate scenarios developed for the California Climate Change Assessment. Using these local projections, SMUD translates those projections into potential climate change impacts on utility services, operations, and infrastructure in the region, as shown in Table 2.[4]

**Table 2.** Potential effects to SMUD infrastructure and operations.[5]

| Impact Category | Potential Effects to SMUD Infrastructure and Operations |
|---|---|
| Ambient Temperatures | • More extreme summertime high temperature events, including daytime and nighttime heat waves<br>• Increased warm season electrical load and peak demand<br>• Reduced thermal and hydroelectric generation<br>• Extreme temperature and variability impacts on system reliability<br>• Increasingly severe "one-in-ten" heat storms effects on overall system reliability<br>• Less efficient operation of transmission and distribution systems, including decreases in facility ratings and loss of operating life |
| Wildfires | • Projected increase in wildfire frequency and intensity<br>• Potential wildfire impacts to transmission and out-of-district generation sources |
| Wind Patterns | • Increases or decreases in wind energy production and timing<br>• Increases or decreases in delta breeze cooling capacity |
| Regional Hydrology | • Effects of changes in temperature and precipitation on snowpack in the Sierra Nevada mountains<br>• Changes in timing and volumes of streamflow and impacts on hydroelectric capacity |
| Flooding | • Sacramento flood threats<br>• Localized impacts on electricity infrastructure<br>• Indirect impacts on gas transmission infrastructure in the San Francisco Bay Delta region |

## INTERACTIVE AND SOFTWARE TOOLS

Interactive tools are available to provide, clarify, and explain climate projection data. Some tools are regional, and many rely only on a subset of the available climate data. Consequently, users of interactive tools should explore multiple options and not simply rely on a single tool. Listed below are some examples of resilience planning tools and informational resources:

**Argonne Resilient Infrastructure Tools:** Argonne National Laboratory's Resilient Infrastructure Initiative focuses on delivering science and technology to enable the resilient design of future infrastructure systems, thereby reducing risk to lives and property. Argonne offers a wide range of resiliency-related capabilities, tools, techniques and engineering methods to optimize interdependencies and respond to rapidly changing needs. Argonne National Laboratory's tools can be made available through the Lab's Federal Technical Assistance Programs. See http://www.anl.gov/egs/group/resilient-infrastructure/resilient-infrastructure-capabilities.

**Cal-Adapt:** The California Energy Commission has combined a large number of climate projections into simple, interactive web maps of California. These maps display the geospatial distribution of changes to climate factors, allowing users to identify potential climate change risks (including temperature, snowpack, sea-level rise, and wildfire probability) in specific geographic areas throughout the state. Climate data in the tool is drawn from multiple university, government, and NGO sources. See http://cal-adapt.org/.

**Cities Impacts and Adaptation Tool (CIAT):** The University of Michigan Climate Center hosts CIAT, a tool that supplies localized climate projections for cities across the Midwest. CIAT provides mid-century annual and seasonal climate projections for temperature and precipitation as well as an interactive map of climate projections. See http://graham-maps.miserver.it.umich.edu/ciat/home.xhtml.

**DOE Sea-Level Rise (SLR) and Storm Surge Effects on Energy Assets:** DOE's Office of Electricity Delivery and Energy Reliability has produced a mapping tool that allows users to view the major energy assets and coastal flooding risks along U.S. coastlines in 10 major metropolitan areas. The tool includes flooding threats from both SLR and hurricane-associated storm surge, but does not include wave threats. See http://energy-oe.maps.arcgis.com/apps/MapSeries/index.html?appid=244e96e24b5a47d28414b3c960198625.

**Hurricane-Induced Coastal Erosion Hazards:** The USGS coastal erosion hazard mapping project displays the probabilities of hurricane-induced erosion at a high level of geographic detail for segments of the Atlantic and Gulf coastlines. The tool is helpful for estimating the vulnerability coastal areas to wave collision, overwash, and inundation as the result of Category 1–5 hurricanes. See http://olga.er.usgs.gov/hurricane_erosion_hazards/.

**National Climate Change Viewer (NCCV):** The USGS NCCV interactive tool allows users to view both graphical and tabular presentations of high-resolution, downscaled CMIP5 projections for four different scenarios—based on individual models or using the average across all models. Users can view temperature and hydrological projections at the state, county, or watershed levels. See www.usgs.gov/climate_landuse/clu_rd/nccv.asp.

**NOAA Sea-Level Rise (SLR) Viewer**: NOAA's Digital Coast mapping tool of the U.S. coastline includes an interactive feature that displays sea-level rise up to 6 feet above the average highest tides and allows users to identify potential inundation risks. The tool does not account for additional increases in flood stage generated by waves or storm surge. See https://coast.noaa.gov/digitalcoast/tools/slr.

**U.S. Climate Resilience Toolkit**: The toolkit is an online resource designed to help people find and use tools, information, and subject matter expertise to understand and manage their climate-related risks and opportunities, permitting them to build climate resilience. The Toolkit provides authoritative, easily accessible, usable, and timely data, information, and decision-support tools on climate preparedness and resilience. See https://toolkit.climate.gov/topics/energy-supply-and-use.

**Climate Explorer**: The Climate Explorer is the central tool that was built to accompany the U.S. Climate Resilience Toolkit, offering customizable graphs and maps of observed and projected temperature, precipitation, and related climate variables for every county in the contiguous United States. Decision makers can compare climate projections based on two scenarios of future climate conditions and plan according to their tolerance for risk and the timeframe of their decisions. See https://toolkit.climate.gov/climate-explorer2/.

**Case Study: Southern California Edison's Adaptation Planning Tool**

Southern California Edison (SCE) is the country's second-largest utility in terms of the number of customers served. It operates in California's second-largest service territory, stretching from the Owens Valley in the north to Riverside and Orange Counties in the south. Building on the climate change assessment resources and tools available in California, SCE developed an Adaptation Planning Tool that uses time-series geospatial datasets to display climate hazard projections across SCE's 50,000-square-mile service territory. The tool uses datasets provided by Cal-Adapt for multiple hazards, including the following:

- Average, maximum, and minimum temperatures
- Precipitation, snowpack, and runoff
- Sea-level rise
- Wind
- Wildfire

The tool shows the locations of SCE facilities and infrastructure, including generation facilities, substations, and transmission and distribution lines. Tool inputs are based on the CCSM3.0 climate model using an A2 scenario for 2030, 2050, and 2085, but the tool also allows the use of alternative, downscaled geospatial climate inputs, such as those from other GCMs. By combining asset locations with future climate hazards in geospatial analysis software, SCE can identify potential climate impacts on assets across its diverse territory and design unique response measures appropriate for individual facilities or systems. Figure 4 shows the Adaptation Planning Tool displaying fire risk.[6]

**Figure 4.** SCE's Adaptation Planning Tool allows planners to identify the projected change to future fire risk across the utility's service territory.[7]

## CREATE NEW CLIMATE PROJECTIONS

In cases where existing climate change projections are inadequate to support a utility's response to key goals and motivations, new climate change projections may be generated by consulting a climate modeling group. The best climate models are computationally complex, but many modeling groups are available (typically academic institutions or consulting firms) and might work as partners to run custom simulations.

Advantages of this approach include the ability to assess custom-defined scenarios and conduct uncertainty studies around hazards specifically tailored to a utility's needs. Given the complexity of climate modeling, however, custom studies can be both costly and time-consuming. Moreover, custom scenarios typically are based on a single model, so users will not be able to examine multiple projections that provide a range of outputs, as with the combinations of models used in ensemble studies.

## CONSIDERATIONS FOR EVALUATING CLIMATE HAZARDS OVER LARGE AREAS

Utilities with service areas spread over multiple states or regions will likely need to consider differences in projected climate change hazards, resulting in widely different vulnerabilities. Conducting a vulnerability analysis across large geographic areas or multiple areas requires additional methodological considerations for each type of climate input. Most importantly, a consistent set of climate change scenarios, models, and other assumptions should be considered when comparing projected hazards across the geographic scope of analysis. Choices about which scenarios, models, assessment literature, or other climate inputs to use may be influenced by stakeholder preference, risk tolerance of decision makers, available resources, or other factors that vary by region. Stakeholder engagement can clarify the purpose of the assessment and inform these choices. Analyses over large areas are likely to encounter diverse preferences and risk tolerances among stakeholders, and the choice of scenarios may need to balance tradeoffs between the total number of scenarios and available resources.

## 2.2 DEVELOP AN INVENTORY OF ASSETS AND IDENTIFY OPERATIONS

One critical input to the vulnerability assessment is an inventory of the assets and operations that could be affected by climate-related threats. Identifying, characterizing, and inventorying a utility's assets and operations will provide useful insights on the various ways in which climate impacts may disrupt services and how best to prioritize and implement operational resilience measures.

### TYPES OF ASSETS AND IMPORTANT ATTRIBUTES

An ideal inventory of assets will include the type of information that will be useful both in evaluating the vulnerabilities of assets to climate impacts (e.g., height above average high water levels) and in deciding which potential resilience measures to pursue. The exact set of attributes to record will depend on the types of assets, the climate hazards being examined, the type of analysis being conducted, and the utility's need to integrate climate vulnerabilities into its risk management framework. Example asset categories are provided in Table 3. Example asset attributes include the following:

- Age of asset, design lifetime, and lifetime of the corresponding licenses or permits

- Geographical location, including lot and structure boundaries

- Elevation, including lot elevation, lowest exterior wall, or lowest floor, and any relevant flood protection

- Current/historical performance and condition, and design operating conditions (including load, temperature, time, etc.)

- Replacement cost, outage cost ($/hour)

- Repair/maintenance schedule and costs

- Vegetation survey

**Table 3.** List of potentially vulnerable types of assets.

| Electric Power Sector Category | Electricity Asset Type |
|---|---|
| Generation | • Steam generator and turbine units<br>• Generator cooling water intake systems<br>• Water filtration and handling equipment<br>• Electrical substation<br>• Back-up power supply sources<br>• Fuel handling and storage systems<br>• Distributed generation units (like solar, back-up diesel units) |
| Transmission | • Long-distance transmission wires and towers<br>• Station control buildings<br>• Substation assets:<br>  – Circuit breakers<br>  – Grounding structure<br>  – Transformers and cooling systems<br>  – Bus bars<br>  – Underground cables<br>  – Protection/control equipment |
| Distribution | • Distribution transformers<br>• Feeder circuits<br>• Switches<br>• Primary circuits<br>• Electric poles |
| General | • Headquarters and operations centers<br>• Fleet storage and service centers<br>• Roads, parking lots, and right-of-way access routes |

## TYPES OF OPERATIONS AND IMPORTANT ATTRIBUTES

As with assets, identifying, characterizing, and inventorying a utility's operations can be an essential step to understanding how climate impacts may disrupt services and how to best prioritize and implement operational resilience measures. Although each utility may have its unique organizational aspects, most utilities share functions similar to the operations identified in Table 4. Regardless of how the operations are categorized, important attributes to inventory for each operation may include the following:

- Number and types of staff, including employees and contractors
- Locations of critical facilities and staff
- Critical equipment, including numbers, types, and locations
- Communications methods and supporting systems
- Data and forecasts necessary for scheduling, planning, and conducting maintenance operations
- Deployment costs for various maintenance functions

Table 4. List of potentially vulnerable utility operations.

| Electric Power Sector Category | Electricity Operation Type |
|---|---|
| Generation | • Fuel procurement<br>• Emissions measurement and verification<br>• Scheduling and control operations<br>• Water handling system maintenance<br>• Boiler, steam system, and turbine maintenance<br>• Switching yard and electrical systems maintenance<br>• Security operations<br>• Planning and construction operations<br>• Facility staffing and accommodations |
| Transmission | • Vegetation management, tower and facilities maintenance<br>• Scheduling & control operations<br>• Emergency response operations<br>• Substation and transformer maintenance<br>• Planning and construction operations<br>• Access route maintenance |
| Distribution | • Vegetation management, pole and facilities maintenance<br>• Distribution line, transformers, and substation maintenance<br>• Emergency response operations<br>• Control operations<br>• Planning and construction operations |
| General | • Headquarters and administrative operations<br>• Capital and resource-adequacy planning<br>• Fleet maintenance<br>• Public and internal communications operations |

**Case Study: Hoosier Energy Identifies Critical Facilities and Business Functions**

Hoosier Energy, a generation and transmission cooperative serving Indiana and Illinois, completed a vulnerability and risk assessment that identifies their most important assets and operations and examines the effects that climate-driven threats may have on them. To identify priority assets, the assessment relies on the following definition for 'critical facilities,' determined in response to NERC guidelines:

> A critical facility may be defined as any facility or combination of facilities that, if severely damaged or destroyed, would have a significant impact on the ability to serve large quantities of customers for an extended period of time, would have a detrimental impact on the reliability or operability of the electric grid, or would cause significant risk to public health and safety.[8]

Using this definition, the assessment identified a set of eight facilities, including four generating stations (and associated switching yards), three substations, and one additional structure.

To identify critical business functions, Hoosier relied on interviews with department managers and other key staff. The interviewees were asked to identify business functions in their areas of responsibility that "would be vital to the continued operation of Hoosier Energy in the event that normal business activities were interrupted by some catastrophic event."[9] The resulting critical business functions are organized into 18 high-level activities with specified locations, key services and assets, and minimum levels of staffing.[10]

## SUPPLIER AND CONNECTED-SECTOR INFRASTRUCTURE AND OPERATIONS

Beyond the fence line of utility-owned assets and operations, many important systems are vulnerable to climate change hazards that could impair or disrupt utility services. Vulnerable connected sectors include any of a utility's suppliers, customers, or other entities with interconnected physical or operational systems. Figure 5 shows the relationship of a utility's infrastructure to other systems and customers.

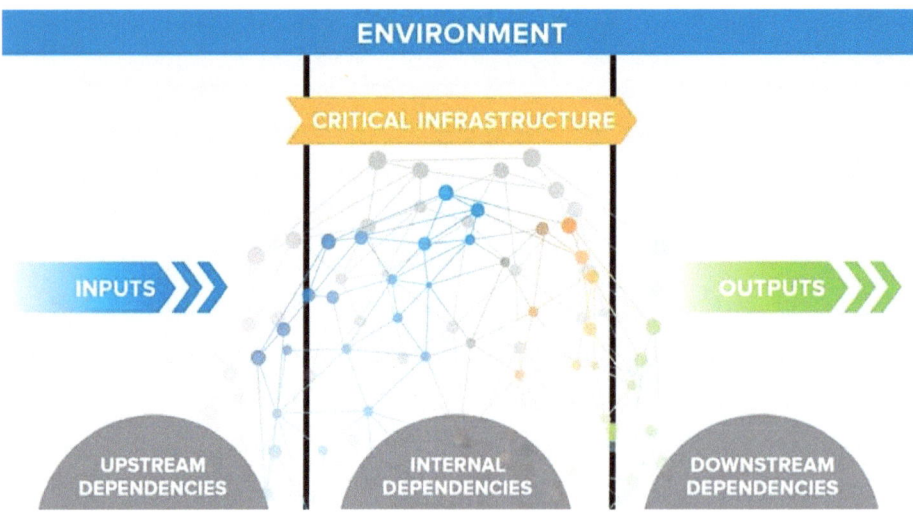

**Figure 5.** Supplier and connected-sector infrastructure.[11]

Vulnerable supplier sectors may include any of the following:

- Generators and transmission equipment providers
- Fuel production systems
- Fuel delivery systems
- Telecommunications systems
- Transportation infrastructure

A utility's customers can include end users (residential and commercial ratepayers) as well as other electricity retailers (using a utility's distribution equipment). If a utility owns generation assets, its customers may also include other utilities or institutional end users. Utility vulnerabilities stemming from climate or extreme weather impacts on a utility's customers are somewhat limited, and primarily include the potential for large or sudden changes in total or peak demand. If the scope of an assessment includes supplier vulnerabilities, utilities may consider capturing the following attributes for each sector:

- Goods and services provided to utility, including vendor/provider, location, and schedule
- Historical range of prices, including any rapid changes in price
- Existing implicit or explicit contingency plans for disruption, including replacement cost
- Redundancy or resilience of supplier infrastructure/networks

## 2.3 REVISIT SCOPE – ITERATIVE PROCESS

After reviewing the climate change resources available and identifying which assets and operations should be considered as part of a vulnerability assessment, it may be helpful to revisit the assessment scope defined in Chapter 1. If new resources, data, or tools were discovered during this step and might expedite or provide further details for the analysis, it may be feasible to consider a wider or more comprehensive scope. Conversely, if reliable data necessary for a complete analysis is not available, it may be necessary to moderate the scope.

# CHAPTER 2 REFERENCES

[1] LLNL (Lawrence Livermore National Laboratory). 2014. *Downscaled CMIP3 and CMIP5 Climate and Hydrology Projections,* UCRL-WEB-236256. Lawrence Livermore National Laboratory. Last modified July 25. http://gdo-dcp.ucllnl.org/downscaled_cmip_projections/dcpInterface.html#Welcome.

[2] Kotamarthi, R., L. Mearns, K. Hayhoe, C.L. Castro, and D. Wuebbles. 2016. Use of Climate Information for Decision-Making and Impacts Research: State of Our Understanding. U.S. Department of Defense, Strategic Environmental Research and Development Program, Washington, DC. March. https://www.researchgate.net/publication/304204988_Use_of_Climate_Information_for_Decision-Making_and_Impacts_Research_State_of_Our_Understanding.

[3] Kotamarthi, R., L. Mearns, K. Hayhoe, C.L. Castro, and D. Wuebbles. 2016. Use of Climate Information for Decision-Making and Impacts Research: State of Our Understanding. U.S. Department of Defense, Strategic Environmental Research and Development Program, Washington, DC. March. https://www.researchgate.net/publication/304204988_Use_of_Climate_Information_for_Decision-Making_and_Impacts_Research_State_of_Our_Understanding.

[4] SMUD (Sacramento Municipal Utilities District). 2012. *Climate Readiness Strategy: Overview and Summary Findings.* Sacramento, CA: SMUD. November. www.hackingsolar.org/library/images/a/a5/SMUD_Climate_Readiness_Report_2012.pdf.

[5] SMUD (Sacramento Municipal Utilities District). 2012. *Climate Readiness Strategy: Overview and Summary Findings.* Sacramento, CA: SMUD. November. www.hackingsolar.org/library/images/a/a5/SMUD_Climate_Readiness_Report_2012.pdf.

[6] SCE (Southern California Edison) 2016. *Progress Update to the Department of Energy.* Rosemead, CA: SCE. February.

[7] SCE (Southern California Edison) 2016. *Progress Update to the Department of Energy.* Rosemead, CA: SCE. February.

[8] NERC (North American Electric Reliability Corporation). 2011. Security Guidelines for the Electricity Sector: Physical Security—Substations. Atlanta, GA: NERC. October. http://www.nerc.com/docs/cip/sgwg/Physical%20Security%20Guideline%202011-10-21%20Formatted.pdf.

[9] Hoosier Energy. 2016. *Vulnerability and Risk Assessment.* Bloomington, IN: Hoosier Energy. February.

[10] Hoosier Energy. 2016. *Vulnerability and Risk Assessment.* Bloomington, IN: Hoosier Energy. February.

[11] Petit, F., D. Verner, D. Brannegan, W. Buehring, D. Dickinson, K. Guziel, R. Haffenden, J. Phillips, and J. Peerenboom. 2015. *Analysis of Critical Infrastructure Dependencies and Interdependencies,* Argonne National Laboratory, Risk and Infrastructure Science Center, Global Security Sciences Division, ANL/GSS-15/4, Argonne, Ill., USA, http://www.ipd.anl.gov/anlpubs/2015/06/111906.pdf.

## 3. DETERMINE EXPOSURE OF ASSETS AND OPERATIONS TO CLIMATE HAZARDS

After a utility has identified appropriate information resources for climate change projections and for the range of its assets or operations to be included in the assessment (Chapter 2), a utility can consider the potential effects of climate change. Identifying vulnerabilities requires an evaluation of both the exposure of assets and operations to potential climate hazards and an estimation of the likelihood or extent of damage or disruption if a hazard occurs. It is important to recognize that not all assets identified as potentially exposed will necessarily be vulnerable, due to a number of factors discussed in this chapter.

**Step 1: Scope the resilience plan**

**Step 2: Develop inputs for vulnerability assessment**

**Step 3: Determine exposure of assets and operations**

**Step 4: Estimate consequences of climate change impacts**

**Step 5: Assess vulnerabilities**

**Step 6: Identify and assess resilience measures**

**Step 7: Build portfolio of resilience measures**

**Step 8: Monitor, evaluate, and reassess**

**Objectives:**

❑ Identify types of climate change hazards and associated electricity sector vulnerabilities

❑ Understand and identify methods for assessing operational and asset vulnerabilities, including screening and detailed analyses

❑ Understand the scaling considerations associated with wide-scale climate hazards

❑ Consider means to determine the likelihood or severity of damage or disruption, given a climate event

### 3.1 CLIMATE CHANGE HAZARDS AND ELECTRICITY SECTOR VULNERABILITIES

Identifying the climate vulnerabilities of assets and operations requires a detailed knowledge of projected climate change hazards and the factors affecting the likelihood of each potential impact (e.g., region, geography, and hydrology, among others). These potential impacts should then be evaluated in terms of the utility's own assets and operations, considering specific locations and other relevant attributes.

One can gain a basic understanding of the various types of climate hazards by consulting existing resources that inventory the potential impacts and the relevant vulnerabilities of electric utilities. Several reports detail climate change vulnerabilities relevant to the energy sector:

> **USGCRP *Third National Climate Assessment* (2014):** Published in 2014, the NCA provides high-level descriptions of climate hazards relevant to the energy sector. The report's region- and sector-specific chapters provide examples and some quantitative details of projected climate hazards. The report also includes some discussion of potential adaptation options and activities underway. DOE produced a technical input report on *Climate Change and Energy Supply and Use*, which provides additional detail. Full report: http://nca2014.globalchange.gov/.
> Technical input for Energy Supply and Use: www.esd.ornl.gov/eess/EnergySupplyUse.pdf.

> **Climate Change and the U.S. Energy Sector: Regional Vulnerabilities and Resilience Solutions (2015):** This report provides detailed and comprehensive accounting of the most significant climate change hazards affecting the energy sector. The report is organized by region to address geographic differences

among energy systems and climate hazards. The report also provides examples of resilience solutions that have been previously implemented. See http://energy.gov/sites/prod/files/2015/10/f27/Regional_Climate_Vulnerabilities_and_Resilience_Solutions_0.pdf.

***U.S. Energy Sector Vulnerabilities to Climate Change and Extreme Weather* (2013):** This report explores different types of climate vulnerabilities experienced by the energy sector, grouped into three major areas: temperature-related impacts; water availability-related impacts; and impacts related to storms, flooding, and sea-level rise. See http://energy.gov/sites/prod/files/2013/07/f2/20130716-Energy%20Sector%20Vulnerabilities%20Report.pdf.

**Effect of Sea Level Rise on Energy Infrastructure in Four Major Metropolitan Areas (2014):** As a pilot analysis, this study applies a flexible and scalable methodology to identify energy facilities exposed to rising sea levels through 2100. The study examines sea level rise exposure for energy assets in Miami, Los Angeles, New York, and Houston. See http://energy.gov/oe/downloads/effect-sea-level-rise-energy-infrastructure-four-major-metropolitan-areas-september.

Multiple approaches are available to compare projected climate hazards asset and operational attributes in order to identify potential vulnerabilities. This guide presents two basic approaches for identifying and classifying vulnerable assets:

- Screening analysis, which examine the effects of a single climate hazard on a large number of facilities, assets, or operations
- Detailed analysis, which looks at the potential effects of multiple projected climate hazards on individual facilities, systems, or operations.

These two approaches are discussed in detail in the following sections.

## SCREENING ANALYSIS FOR LARGE SETS OF ASSETS

For systems with large numbers of assets or for operations dispersed across a large geographical area, a screening analysis is a useful way to identify all potentially vulnerable locations or to characterize the scale of a potential vulnerability. This section describes the use of screening analyses to identify which assets or locations will be vulnerable to a specific, quantified climate hazard. A screening analysis may be completed for separate climate hazards, but the approach is best used for cases in which there are regional variations either in the projected climate hazards (e.g., monthly precipitation or high temperature) or in the attributes of a utility's assets and operations (e.g., height above sea level or safe operating temperature). Regional variations in either could affect their vulnerability to potential impacts.

A screening analysis typically involves identifying a critical threshold for a specific climate screening parameter. These thresholds are simply values above or below which the likelihood of a climate impact is considered sufficient to render the asset or operation vulnerable. Critical thresholds should be based on the asset and operational attributes identified in Chapter 2. Examples include the following:

- Historical operating conditions associated with damage, accelerated wear, increased costs, or service interruption/disruption

- Design parameters or regulated operating parameters
- Quantifiable physical characteristics of assets or facilities

For some climate hazards, a threshold indicates a clear point at which damage or disruption could occur (e.g., intake water temperatures above which a nuclear power plant cannot operate). For other climate hazards or potentially vulnerable assets or operations, a threshold can be set as a point along an increasing slope of likelihood that the asset will suffer a significant cost or impact. In setting thresholds, a planner tries to identify the point above which the risk of impact is great enough to qualify as a vulnerability.[c] Tables 5 and 6 list assets and operations, respectively, that may be vulnerable to specified climate impacts and provide examples of some appropriate screening parameters. Thresholds are often system or component-specific.

Projected climate hazards and quantified screening parameters can be compared and evaluated by various means, including spreadsheets, statistical software, or Geographic Information System (GIS) packages. A GIS can overlay multiple geo-referenced databases simultaneously and quickly compare projected changes in climate parameters to a variety of other relevant datasets.[1,2]

Table 5. Sample assets and climate hazard screening parameters.

| Climate Change Hazard | Sample Vulnerable Assets | Sample Vulnerabilities | Screening Parameter |
|---|---|---|---|
| Rising sea levels | • Power plants/ switching yards <br> • Structures, parking lots, operations facilities | • Periodic or permanent inundation <br> • Increased risk of storm surge flooding | • Structure elevation above sea level |
| Increasing temperatures; higher peak temperatures; longer, more frequent heat waves | • Transmission & distribution transformers | • Reduced transformer loading capacity <br> • Accelerated breakdown of transformer insulation | • Transformer and cooling system safe operating temperatures |
| Increased hurricane-related wind intensity | • Transmission and distribution power poles | • Increased risk of wind damage to power poles <br> • Increased risk of vegetation damage | • Wind speed at which elevated risk of damage may occur, by type of pole |

---

[c] In general, risk tolerances should be presumed to be low for identifying vulnerabilities; greater focus on relative risks will be required while prioritizing response measures later in the process. Quantifying the functional relationship between a screening parameter and cost is discussed further in Chapter 4.

**Table 6.** Sample operations and climate hazard screening parameters.

| Climate Change Hazard | Sample Vulnerable Operations | Sample Vulnerabilities | Screening Parameter |
|---|---|---|---|
| Increasing temperatures, increasing CDDs | • Day-ahead power scheduling<br>• Resource adequacy planning | • Increasing peak electricity demand during hot days | • Daily high temperatures and related peak demand |
| Longer growing season, increased risk of wildfires | • Vegetation management of rights-of-way along transmission and distribution lines | • Accelerated inspection and trimming schedules | • Wildfire frequency |

**Case Study: National Grid Uses GIS Analysis to Screen Vulnerable Substations in Three States**

National Grid operates electric utilities serving over 3 million customers in five U.S. states. In 2013, the company performed a Substation Flood Study to assess vulnerability to flooding for its substations in Massachusetts, New York, and Rhode Island.[3] The assessment used GIS software to overlay the elevations of its substations and substation assets with FEMA-produced Flood Insurance Rate Maps (FIRMs). Although existing FIRMs do not account for climate change projections, National Grid plans to update its assessment as FEMA issues newer FIRMs that consider sea-level rise and enhanced risk of storm surge.

National Grid determined that asset elevation was the appropriate screening parameter for the assessment and conducted a field survey to collect the elevations of substation yards, structure and asset foundations, and key equipment panels. Substations were rated as High- or Medium-risk if located inside the 100-year or 500-year flood zones, respectively, or as Low-risk if located outside of the 500-year flood zone. Equipment within each substation was ranked as High-, Medium-, or Low-risk, depending on the equipment's elevation.

At substations rated at Low- or Medium-risk, National Grid did not take any immediate action. At High-risk substations, National Grid implemented flood avoidance or mitigation measures for any equipment ranked at High-risk (equipment located below the base flood elevation) and implemented measures to make Medium-risk equipment (equipment less than two feet above base flood elevation) flood repairable. These measures include both short-term fixes as well as long-term solutions. Short-term fixes include elevating specific equipment, installing berms or barriers, or installing connections for mobile substations in case of a flood. Long-term solutions include retiring or relocating the substations.[4]

## DETAILED ANALYSIS FOR INDIVIDUAL ASSETS OR FACILITIES

For large assets or for any important assets identified as potentially vulnerable in a screening analysis, it is appropriate to conduct a detailed review of exposure to a climate change hazards. Detailed analyses are also useful for evaluating the vulnerability of utility operations. A detailed analysis involves individual consideration of each asset (or operational) attribute and how projected climate hazards may affect these attributes in the future, and may incorporate detailed historical data and custom modeling. Like screening analyses, a detailed analysis should use quantitative measures whenever possible to evaluate potential vulnerabilities (e.g., comparison of structure elevations with projected storm surge heights), but should consider all climate hazards within the scope of the vulnerabilities assessment.

The objectives of a detailed vulnerabilities analysis are as follows:

- Evaluate potential vulnerabilities to a complete range of projected climate hazards
- Verify vulnerabilities identified via screening analysis
- Examine vulnerabilities that cannot be easily reduced to a single screening parameter, or for which reliable, localized, quantitative projections are not available
- Examine vulnerabilities that may arise from a complex set of climate- and non-climate-related hazards (e.g., sea level rise, increased hurricane intensity, local subsidence, and wave action)
- Consider probabilistic information regarding the likelihood of certain low-frequency, high impact climate or extreme weather events or outcomes, wherever possible

## EXAMPLES OF DETAILED ANALYSES

**Thermoelectric Power Plants Located on Rivers:** The temperature of river water is a critical factor in the safe, efficient, and environmentally responsible operation of thermoelectric power plants. When water temperatures become too high, the efficiency and output of a steam-cycle plant falls—often during extreme heat events when electricity is needed most. In addition, elevated water temperatures reduce the amount of hot water that power plants can return to the river without exceeding the thermal discharge limits imposed to protect river ecology. River water temperatures are not a standard output of GCMs and are not typically included in climate projection resources like those identified in Chapter 2; however, a detailed analysis of potential vulnerabilities at a thermoelectric power plant could use historical air and water temperatures and precipitation records to generate a functional relationship between those climate inputs and river water temperatures. Using such site-specific derived relationships in combination with future climate projections could help evaluate the vulnerability of a thermoelectric power plant to future increases in air and water temperature.

**Distribution Operations Center Located Near Coastline:** A coastal utility operations center with offices, storage and maintenance facilities, and a large parking area could be screened for potential vulnerabilities to coastal flooding using the elevation of either the centroid or the lowest point of the property. A detailed follow-up analysis should consider the lowest elevation of each structure or piece of equipment, the vulnerabilities of support infrastructure (e.g., entry roads, power systems, telecommunications systems), and any relevant flood protection measure (e.g., floodwalls, elevated or submersible equipment, etc.). Projected climate hazards (including sea level rise and increased storm surge intensity) should be refined by data on local variations in coastal slope, wave height, and local land subsidence. Exposure analysis should also take into account the history of storm impacts in a location and consider how climate projections may affect the probability of storm events in the future.

**Case Study: Exelon Models River Flow Impacts on Braidwood Power Station**

In 2013, Exelon established a Drought Monitoring Task Force to assess existing drought conditions, historical drought impacts, and potential future impacts under climate change scenarios. The Drought Task Force identified the Braidwood Nuclear Power Station as potentially vulnerable to drought impacts. Braidwood is a two-unit nuclear power plant with a nameplate capacity of 2,520 MW. Its recirculating cooling system draws from a 2,500-acre cooling pond fed by the nearby Kankakee River.

During the Midwest drought/heat wave of 2012, Braidwood reached its permitted low-flow river water-withdrawal limit. To better understand Braidwood's vulnerability to low-flow conditions, Exelon arranged a hydrologic study of the Kankakee River watershed. The study identified and analyzed projected climate change impacts using a hydrologic model of the Kankakee Basin. In addition to climate change, the study evaluated future population growth, development, and potential changes in environmental protection regulations. The model ran numerous scenarios, including an increase of 50% in upstream water use during typical low-flow months, and extrapolated to 2040.

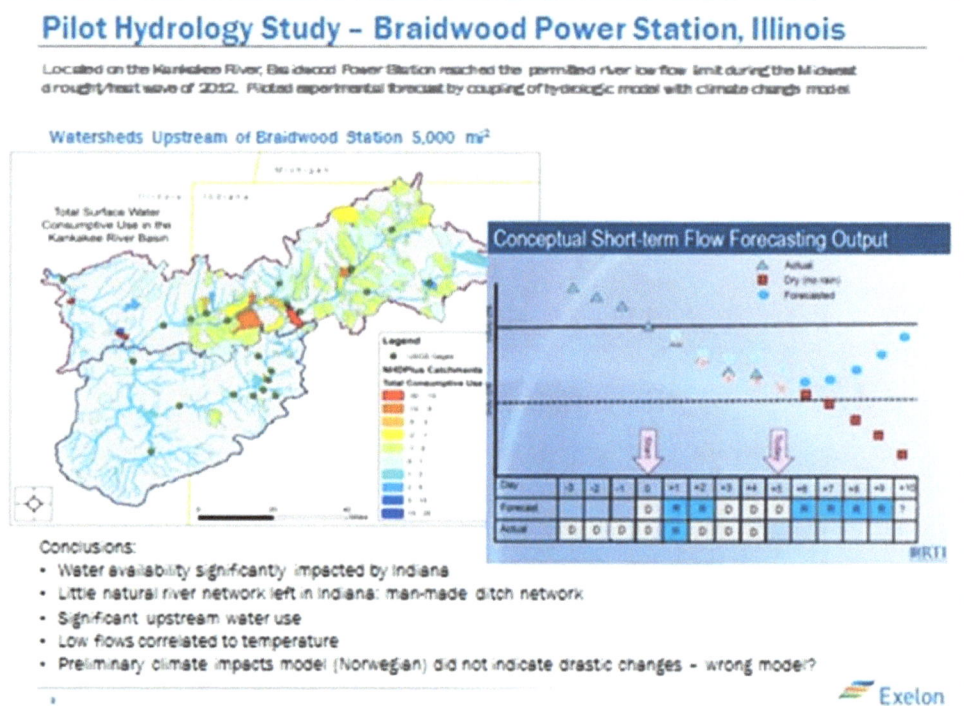

**Figure 6.** Exelon's Pilot Hydrology Study used hydrological models and climate change projections to determine risks associated with falling water levels and rising water temperatures at its Braidwood plant.[5]

As a result of the pilot study, Exelon discovered significant limitations on the ability to model future weather and predict the effects of climate change and other factors on long-term water availability at the local level. Exelon is continuing to pursue cutting-edge research in an effort to better understand potential climate and water impacts. Exelon is also building its internal understanding of river levels and temperatures by installing upstream monitoring systems in the watersheds of rivers used to cool power plants. The resulting data is processed by models run multiple times per day and used for a Daily River Report, which provides upstream river stage and temperature.[6]

**Case Study: San Diego Gas & Electric (SDG&E) Works with Partners to Develop Wildfire Threat Index**

After a series of destructive wildfires related to the Santa Ana winds in southern California, SDG&E collaborated with researchers at the U.S. Forest Service (USFS) and UCLA to conduct a detailed analysis of Santa Ana winds and their influence on southern California wildfires. Following over three years of analysis, the team produced the Santa Ana Wildfire Threat Index (SAWTI). SAWTI provides a six-day forecast of wildfire threat based on meteorological and fuel moisture data. SAWTI measures the likelihood of large fires (specifically, the probability of a fire reaching or exceeding 250 acres in size), and groups the index into five categories: No Rating, Marginal, Moderate, High, and Extreme.

SDG&E and firefighting agencies use SAWTI forecasts to anticipate potentially damaging fires and allocate shared response resources efficiently. SDG&E has also used SAWTI to examine in detail the potential increase in wildfire vulnerabilities to its assets and operations associated with changing temperatures and precipitation.[7,8]

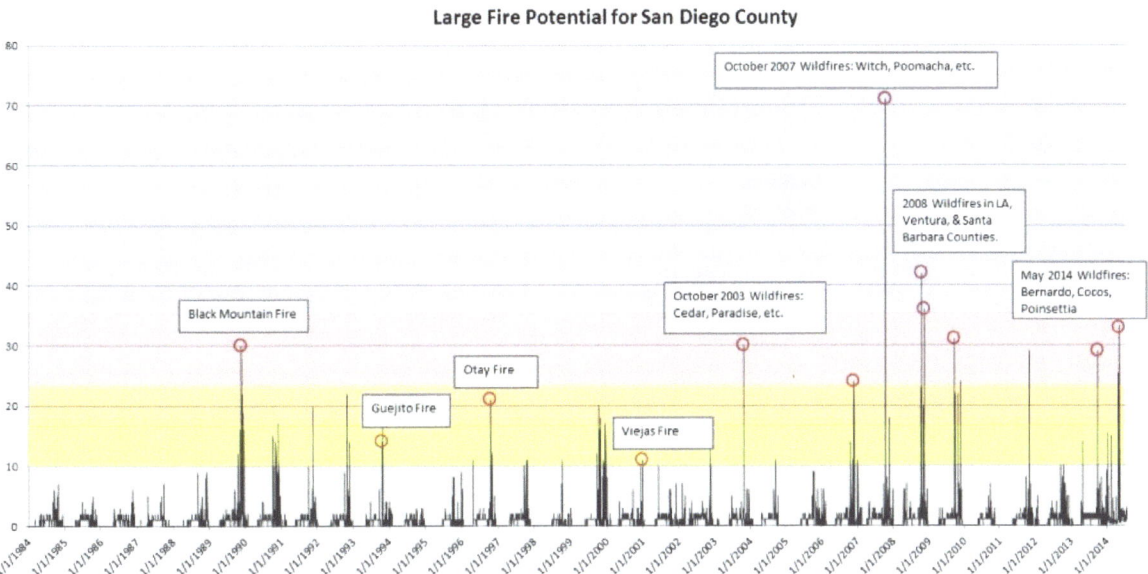

**Large Fire Potential for San Diego County**

**Figure 7. Historical** SAWTI values for 1984–2014. Yellow band is Marginal threat, orange indicates Moderate, red indicates High, and purple indicates Extreme threat; red circles indicate major fires.[9]

## SCALING CONSIDERATIONS

Climate change and extreme weather hazards have the potential to damage diverse types of energy and other essential infrastructure and to disrupt critical systems over a large geographical area. Infrastructure damage at this scale can cause extended disruptions and require a longer recovery time than the cumulative total if each asset were restored individually. Complications may stem from disruptions to other infrastructure and systems (such as fuel deliveries, telecommunications, etc.), a shortfall of response capability (including mutual assistance capacity), and the need to black start generation and coordinate grid recovery.

A screening analysis can provide insight on the potential scale of vulnerability to a widespread climate hazard or extreme weather event across an entire power system. For example, while an individual transformer may be vulnerable to elevated peak temperatures, a screening analysis can indicate the total number of transformers that may be forced to reduce loading during an extreme heat event.

## SUPPLIER AND CONNECTED-INFRASTRUCTURE VULNERABILITIES

In identifying potentially exposed assets and operations, planners should also consider vulnerabilities stemming from climate impacts on suppliers, customers, and other connected infrastructure. Connected infrastructure can include fuel suppliers, telecommunications providers, and transmission operators, among others. Examples of connected infrastructure are shown in Figure 8. Supplier vulnerabilities can affect utility operations by causing shortages in fuel, critical equipment, or services. For example, internet connectivity and other communication systems are essential for some electric sector operations and sustained communication failures could lead to power system disruptions. More-detailed analyses of supplier and connected-infrastructure vulnerabilities may require network analysis to model the response of the power system to climate threats and evaluate the sensitivity of the grid to outages among different utilities and connected infrastructure. Modeling can simulate the operation of the grid under normal and stressed conditions, such as the unplanned loss of transmission facilities.

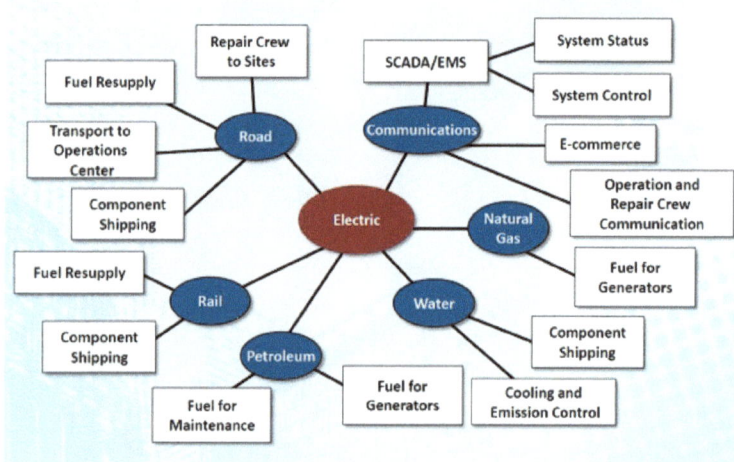

**Figure 8.** Electric power dependencies across sectors.[10]

A utility may decide to collaborate with suppliers, customers, or connected infrastructure owners or operators to identify shared vulnerabilities. Utilities should look for opportunities to collaborate and allow sharing of data, common scenarios, assumptions, and analysis methodologies with both suppliers and customers.

If collaboration with suppliers is not possible, quantitative analysis of connected-infrastructure vulnerabilities will require supplemental or proxy data. One approach is to use publicly available data about suppliers and connected infrastructure in a screening analysis. This method may involve assuming industry-standard operating conditions (e.g., based on published averages) and matching the locations of supplier facilities to public topographic databases (e.g., USGS baseline elevation data).[d] Any quantitative analysis performed on data derived from these types of assumptions must reflect the uncertainty involved in the estimates.

---

[d] USGS maintains *The National Map*, a composite database consisting of multiple geospatial datasets, including topographic and 3D Elevation maps. The elevation layer of the *Map* includes both bare-earth maps called the National Elevation Dataset, as well as the newly operational 3D Elevation Program that includes the locations and elevations of surface structures. *National Map* data can be accessed via the USGS website: http://nationalmap.gov/index.html.

Another approach to addressing supplier vulnerabilities is to examine the potential effects of losing access to the goods or services provided by suppliers and consider these effects as another type of "climate impact." Assets or operations should be considered vulnerable to these impacts if a specified climate hazard could credibly disrupt delivery, although defining threshold parameters will require assumptions about the vulnerability of connected infrastructure.

## 3.2 LIKELIHOOD OR SEVERITY OF DAMAGE OR DISRUPTION

When assessing the vulnerabilities of assets or operations, one should consider the actual repercussions of the climate or extreme weather impact on those assets. In some cases, a climate event can occur without significantly damaging or disrupting the equipment, systems, or operations. In other cases, the severity of damage or disruption may depend on multiple factors, including the intensity of a climate impact, attributes of the asset or operation, or existing protection. In the event of a storm, flood, heat wave, or other climate event, the likelihood or potential severity of damage or disruption is not identical for all assets and operations, or for all climate events. For example, during a heat wave, transmission line outages are made more likely by elevated line temperatures due to increased demand. However, the occurrence of a transmission outage involves multiple and often highly localized factors, such as the local temperature, age and condition of equipment, intermittent wind speed and the presence of overgrown vegetation. Attempts to quantify the effect of climate-driven extreme weather hazards such as heat waves on transmission line outages should consider both the probability of occurrence as well as the probability of severity.

The sensitivity of an asset to a potential climate event depends on both the type and severity of the event (e.g., the force of a wave or temperature during a heat wave) and the type, configuration, or attributes of the asset or operation itself (e.g., the physical resilience of a power pole to increased wind speeds or wave force). For example, during high winds, the probability that any individual power pole will be blown down is greater than zero but less than certain. Assessing the likelihood of damage or disruption will help planners effectively prioritize resilience measures (prioritizing resilience measures is discussed further in Chapter 7).

If the vulnerabilities assessment is using probabilistic estimates of climate parameters or climate hazards, quantitative probabilities of damage or disruption can be developed. However, in most cases, it is difficult to project the likelihood or severity of damage or disruption given the occurrence of a climate or extreme weather impact, so quantifying probabilities is necessarily a process of developing a "best estimate." A variety of factors and sources, including expert opinion, design standards, and post-event reports, can be used in estimating the probability of threats, damage, or disruption from a particular climate event.

> **Expert judgement or elicitation:** In many cases, experts can serve as a primary source of information regarding asset sensitivity. Subject matter experts who understand the assets and operations of utilities as well as the implications of climate events can offer useful judgments about the probabilities that adverse events will damage or disrupt those assets. The success of expert opinion hinges on the well-orchestrated interplay between the right subject matter experts using the right information (or the information available) and analysts who apply appropriate methods to judge event likelihoods and draw the correct inferences from the expert opinions.[11] Although few examples exist where this method has been applied in the area of climate-related risk, expert judgements are best used to estimate the probabilities of events for which historical records are scant or absent. An extensive body of literature exists on the design and proper use of expert elicitation.[12]

**Design standards:** Knowledge of the design and construction of infrastructure can enhance understanding of the sensitivity of many types of assets to climate events. Baseline NESC standards and utility standards may provide information about the weather-related thresholds that certain assets are known to meet without issue, or with predictable failure rates. However, standards may not address required performance in the face of changing climate or extreme weather conditions (e.g., if the frequency or duration of extreme weather events increases). Design criteria vary widely for some components and elements of the energy system. In many cases, existing infrastructures pre-date modern codes that provide criteria to calculate and apply weather-related stressors.[13]

**Infrastructure age:** The age of electricity system assets can also influence asset sensitivity to climate events. In some cases, information on the age of electricity assets may be limited, beyond central station generation assets.[14,15] Older systems are generally more sensitive to damage or disruption from climate hazards and extreme weather.

**Past events:** Historical storm reports or post-event reports can provide useful information on the sensitivity of assets. These reports, which are now appearing more frequently, may provide insights on the types of assets that failed and under what conditions (e.g., DOE's report *Comparing the Impacts of Northeast Hurricanes on Energy Infrastructure*,[16] or New York City's *A Stronger, More Resilient New York*[17]).

**Fragility curves:** Damage functions or fragility curves may be available for some types of assets. These curves describe the relationship between threat intensity or magnitude and asset damage or degree of impact, based on the sensitivity of the asset and its components.[18] FEMA's Hazus model[e] provides basic damage functions for key types of energy assets generically, which can help with high-level assessments. In addition, insurers or catastrophic modeling companies may have proprietary information on fragility curves for energy assets.

## CHAPTER 3 REFERENCES

[1] Bai, Y., I. Kaneko, H. Kobayashi, K. Kurihara, I. Takayabu, H. Sasaki, and A. Murata. "A Geographic Information System (GIS)-based approach to adaptation to regional climate change: a case study of Okutama-machi, Tokyo, Japan." *Mitigation and Adaptation Strategies for Global Change 19:589.* February 19. http://link.springer.com/article/10.1007/s11027-013-9450-6.

[2] FHWA (Federal Highway Administration). 2011. *Applications of Geographic Information Systems (GIS) for Transportation and Climate Change.* U.S. Department of Transportation, Federal Highway Administration, Washington, DC. August. https://www.gis.fhwa.dot.gov/documents/climate_change_report_aug2011.pdf.

[3] National Grid. 2016. *Climate Change Resilience Planning.* Waltham, MA: National Grid USA. February.

[e] See: www.fema.gov/hazus

[4] National Grid. 2016. *Climate Change Resilience Planning*. Waltham, MA: National Grid USA.  February.

[5] Exelon. 2016. Addressing Climate Change Resiliency: Vulnerabilities Assessment. Chicago, IL: Exelon Corporation.  January.

[6] Exelon. 2016. Addressing Climate Change Resiliency: Vulnerabilities Assessment. Chicago, IL: Exelon Corporation.  January.

[7] SDG&E (San Diego Gas & Electric). 2016. *San Diego Gas & Electric Regional Energy Sector Vulnerabilities and Resilience Strategies.* San Diego, CA: San Diego Gas & Electric. February.

[8] SDG&E. 2015. "Prepared Direct Testimony of Steve Vanderburg on Behalf of San Diego Gas & Electric." *Testimony before California Public Utilities Commission.* San Diego, CA: San Diego Gas & Electric. September 25. https://www.sdge.com/sites/default/files/regulatory/FINAL%20Vanderburg%20Testimony.pdf.

[9] SDG&E. 2015. "Prepared Direct Testimony of Steve Vanderburg on Behalf of San Diego Gas & Electric." *Testimony before California Public Utilities Commission.* San Diego, CA: San Diego Gas & Electric. September 25. https://www.sdge.com/sites/default/files/regulatory/FINAL%20Vanderburg%20Testimony.pdf.

[10] Argonne National Laboratory. 2010. *CI/KR Identification and Cross Sector Interdependencies and Dependencies*. Argonne, IL: Argonne National Laboratory.

[11] INL. 2015. *Simplified Expert Elicitation Guideline for Risk Assessment of Operating Events*. Idaho Falls, ID: INL. https://inldigitallibrary.inl.gov/sti/3310952.pdf.

[12] INL. 2015. *Simplified Expert Elicitation Guideline for Risk Assessment of Operating Events*. Idaho Falls, ID: INL. https://inldigitallibrary.inl.gov/sti/3310952.pdf.

[13] NIST (National Institute of Standards and Technology). 2015. *Community Resilience Planning Guide for Buildings and Infrastructure Systems: Volume II.* U.S. Department of Commerce. http://www.nist.gov/el/resilience/upload/NIST-SP-1190v2.pdf.

[14] Panteli, M., and P. Mancarella. 2015. "Modeling and Evaluating the Resilience of Critical Electrical Power Infrastructure to Extreme Weather Events." *IEEE Systems Journal* PP(99): 1–10. February. http://dx.doi.org/10.1109/JSYST.2015.2389272.

[15] NIST (National Institute of Standards and Technology). 2015. *Community Resilience Planning Guide for Buildings and Infrastructure Systems: Volume II.* U.S. Department of Commerce. http://www.nist.gov/el/resilience/upload/NIST-SP-1190v2.pdf.

[16] DOE (U.S. Department of Energy). 2013. *Comparing the Impacts of Northeast Hurricanes on Energy Infrastructure.* Washington, DC: DOE. April.http://energy.gov/sites/prod/files/2013/04/f0/Northeast%20Storm%20Comparison_FINAL_041513c.pdf.

[17] New York City. 2013. *A Stronger, More Resilient New York*. New York City, NY: City of New York. June. www.nyc.gov/html/sirr/html/report/report.shtml.

[18] Panteli, M., and P. Mancarella. 2015. "Modeling and Evaluating the Resilience of Critical Electrical Power Infrastructure to Extreme Weather Events." *IEEE Systems Journal* PP(99): 1–10. February. http://dx.doi.org/10.1109/JSYST.2015.2389272.

## 4. ESTIMATE CONSEQUENCES OF CLIMATE CHANGE IMPACTS

Chapter 4 describes the process and methods for calculating the costs of climate impacts identified in Chapter 3. For each vulnerable asset or operation, the manifestation of a climate change or extreme weather impact may cause direct, indirect, or induced costs. These costs will vary significantly depending upon which assets or operations are affected, the location and severity of the impacts, and the duration of the disruption. For example, although distribution outages occur more frequently than transmission outages, transmission outages generally lead to much higher costs.[1] This section discusses approaches to estimating these costs, which will be useful in conducting a cost-benefit analysis for resilience measures and in prioritizing responses to climate change.

### 4.1 DIRECT, INDIRECT, AND INDUCED COSTS OF CLIMATE CHANGE IMPACTS

**Step 1: Scope the resilience plan**

**Step 2: Develop inputs for vulnerability assessment**

**Step 3: Determine exposure of assets and operations**

**Step 4: Estimate consequences of climate change impacts**

**Step 5: Assess vulnerabilities**

**Step 6: Identify and assess resilience measures**

**Step 7: Build portfolio of resilience measures**

**Step 8: Monitor, evaluate, and reassess**

**Objectives:**

- Distinguish between direct, indirect, and induced costs of climate impacts

- Recognize importance of the non-linear cost growth of widespread impacts

- Identify example methodologies to quantify the costs of climate impacts

Every climate impact carries potential direct costs, which apply to the affected electric utility (asset owner) and indirect costs, which apply to suppliers, customers, or society. The direct and indirect costs associated with impacts on vulnerable assets and operations will be useful in analyzing the costs and benefits of resilience measures, as discussed in Chapter 7. While induced costs are also discussed here for informational purposes, analytical approaches for quantifying induced costs are not introduced.

Utilities face uncertainty as to whether they will be allowed to recover direct costs after an event. This uncertainty is due to a number of regulatory factors. For example, regulations may prohibit "single issue" ratemaking yet permit a periodic general rate case. If regulations do allow cost recovery to be considered, regulators may still consider whether costs were prudently incurred, whether storm-related costs should be deferred, whether to allow the recovery of carrying costs, and other issues that potentially restrict cost recovery and impede climate resilience investments. As noted earlier, utilities and regulators are increasingly turning to other means beyond the ratepayers to deal with funding resilience investments, including: cost deferral, rate adjustment mechanisms, lost revenue and purchased power adjustments, formula rates, storm reserve accounts, securitization, customer or developer funding/matching contributions, federal funding, and insurance.[2]

## DIRECT COSTS

The direct costs of climate impacts on the electric sector are the economic losses to an electric utility. These losses include all of the additional expenditures and administrative and labor costs associated with responding to outages—the costs of repairing, replacing, or relocating facilities and equipment—and the opportunity costs of lost sales during an outage. Table 7 provides examples of direct costs that may be incurred because of climate change and extreme weather impacts.

**Table 7.** Examples of direct costs of climate change and extreme weather impacts.

| Climate Impact | Direct Cost of Impacts |
|---|---|
| **Nuisance Flooding (Periodic, Temporary)** | • Restoration and repair costs, including parts and labor<br>• Replacement costs for damaged assets, including parts and labor<br>• Administration of restoration and repair activities, including inspections, procurement, and installation/removal of temporary measures like portable substations |
| **Permanent Inundation due to Sea-Level Rise** | • Relocation costs, including property, infrastructure, engineering, and installation<br>• Costs to connect relocated assets and supporting infrastructure<br>• Replacement costs for equipment that cannot be relocated |
| **Extreme Storm Surge Event** | • Restoration and repair costs, including parts and labor<br>• Replacement costs for damaged assets, including parts and labor<br>• Administrative costs |
| **Wildfire** | • Inspection and repair/replacement costs for assets damaged by smoke exposure<br>• Replacement costs for assets damaged by fire |
| **Warmer Temperatures and Extreme Heat Events** | • Restoration costs for outages<br>• Replacement costs for equipment needing earlier replacement |

For all types of direct costs, capital and labor will vary according to region, manufacturer, design specifications, and contract relationships, among other factors. Costs for relocation will also vary by the setting, especially if new land must be acquired. Utilities generally collect but do not publicly report detailed damage estimates from storm events based on asset type and location, so a utility's own personnel and asset databases will often be the best source for relevant cost information. Capital investment plans and rate filings[f] can also provide supplementary information on new asset costs, and relevant information may be available from surveys regarding grid vulnerability and resilience.[3,4,5] Several representative examples of costs are provided on the following pages.

---

[f] Following Hurricane Sandy, ConEd engaged with stakeholders via the Storm Hardening and Resiliency Collaborative to plan a program of resilience-building upgrades across its systems. ConEd's rate cases presented in 2014 and 2015 contain substantial detail on the costs of many of these resilience upgrades. ConEd's rate case docket is located here: http://documents.dps.ny.gov/public/MatterManagement/CaseMaster.aspx?MatterCaseNo=13-E-0030.

For high-level estimates of direct costs from climate impacts on a facility, planners could assume that exposed assets would be damaged beyond repair, and use standard replacement costs for generic asset types. To generate high-level estimates of the direct cost for broad asset replacement, planners can multiply the number of exposed assets by the standard generic asset cost. This total loss scenario does not include other costs, such as relocation. Due to the threat of permanent inundation or foundation damage from sea level rise, riverbank erosion, or other climate and extreme weather risks affecting a facility site, utilities may also consider the costs of relocation, which will be highly context specific. For high-level estimates, information from previous relocation investments may be useful (e.g., property values, connection and infrastructure costs, etc.). However, cost estimates based on a replacement-in-kind approach may not reflect the changes or upgrades necessary to achieve an enhanced level of resilience with new equipment.

For more detailed cost estimates, planners should estimate for each asset the level of damage likely to result from climate impacts. As discussed in Chapter 3, factors affecting the probability of damage or disruption should be considered in developing these estimates. The level of detail in a cost analysis will depend on the utility's needs and motivations, but assumptions should remain consistent throughout the assessment. Site-specific information is required to estimate relocation costs if accuracy is important. When estimating damages, the potential for changes to future demand should also be considered. For example, assets in areas likely to experience permanent inundation due to sea-level rise may see reductions in load due to potentially large population migration, effectively limiting the asset's value.

Any analysis of the projected costs of future climate impacts should consider the timeframe of climate impacts.[g] When evaluating the costs of climate impacts alongside resilience measures, costs should be converted to their Net Present Value (NPV) using appropriate discounting.[6,7,8] To more accurately reflect total costs, planners should consider the number of events an asset is likely to encounter over its service life.

---

[g] Timing of events should be based on information gathered as part of understanding the hazard. See Chapters 2 and 3.

**Case Study: Entergy Uses Custom Quantitative Methods to Estimate Direct Costs of Storm Damage**

Entergy's 2007 Hurricane Hardening Study addresses potential damages from projected hurricane events in the utility's service territory.[9] One of the direct costs faced by the utility included damages to wooden distribution poles located throughout the service area. The direct costs of damage to wood poles can vary significantly based on how many are damaged. To estimate the direct cost of poles damaged by hurricane-force winds, Entergy created a model showing the probability of pole failure based on wind speed and pole type (i.e., wood, concrete, lattice steel, or tubular steel). The model correlates data on wind speed, number of poles exposed, and number of failed poles from previous hurricanes, including Katrina and Rita.

### Wood Pole Damage Ratios by Wind Speed

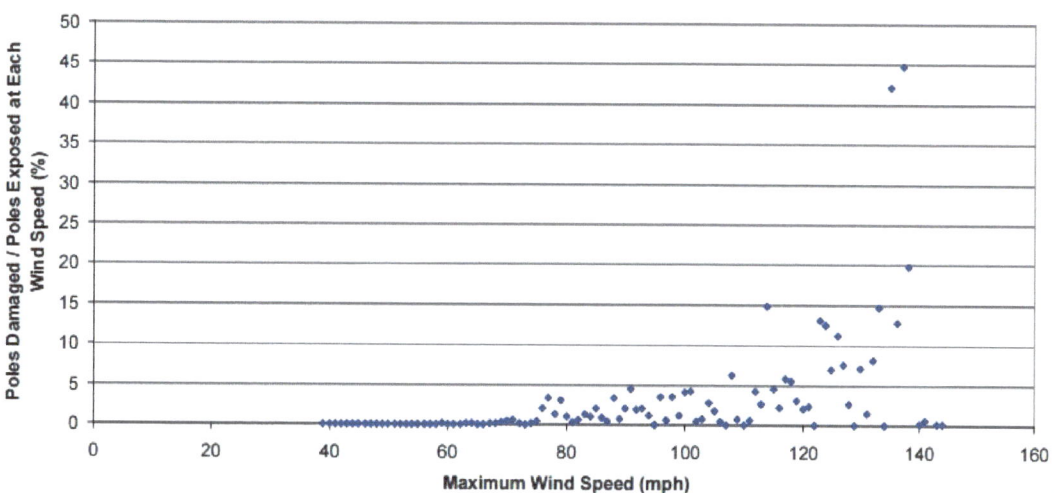

**Figure 9.** Share of exposed wood poles damaged by winds vs. maximum wind speed in Entergy service area.[10]

A best-fit line is generated for each type of pole and used to estimate damages based on projected future wind speeds. To estimate direct costs of projected future damages, the number of projected damaged poles can be multiplied by the replacement cost for each type. The method used by Entergy allows the direct quantitative connection between a climate parameter (wind speed) and a direct cost.[11]

## INDIRECT AND INDUCED COSTS

Indirect and induced costs include those costs experienced by consumers, other companies, or by society as a whole. Primarily, these costs represent the lost value of electrical power during an outage, but they also include any damage to equipment caused by a sudden loss of power, interruptions in interconnected infrastructure, and social costs resulting from a power outage. This section describes all types of indirect costs associated with climate impacts on the electric sector. The analytical methods provided further in this section focus exclusively on indirect costs affecting consumers. Analytical methods to quantify induced social costs, including the costs of electricity outages on other energy sectors are not addressed here. Because induced costs do not directly affect ratepayers, they are typically not useful for establishing the prudence of an investment in utility infrastructure, and are not considered in evaluating the resilience options in Chapter 6, or in cost-benefit analyses described in Chapter 7.

- **Indirect costs**: The loss of electrical service by a utility's customers, including commercial and residential ratepayers and other utilities that purchase power generated by the utility, as well as damage to equipment caused by outages

- **Induced costs**: Costs affecting society (other than costs to a utility's consumers), e.g., companies that have their supplies interrupted, employees losing jobs, etc.

The costs of interruptions in electricity service vary by the class of electricity consumer. Costs to consumers represent the value of the electricity lost, the value of any damages caused by the sudden loss of electricity, and the lost value of reliable electricity delivery. The two latter cost categories are primarily concerns for large users (including commercial, industrial, agricultural, and infrastructure users) rather than for residential users. Reliable electricity is critical for planning large-scale industrial operations and affects the coordination and scheduling of other costly inputs to production (e.g., labor, materials). Sudden outages can damage equipment or cause spoilage of materials; the cost of this damage is included in the category of indirect costs to consumers.

**Table 8.** Examples of indirect and induced costs by consumer class.[12]

| Consumer Class | Indirect Costs to Consumers | Induced Costs to Non-Consumers |
|---|---|---|
| **Residential** | <ul><li>Inconvenience, lost leisure, stress, etc.</li><li>Out-of-pocket costs:<ul><li>– Spoilage</li><li>– Property Damage</li></ul></li><li>Health and safety effects</li></ul> | <ul><li>Costs to other households and firms</li></ul> |
| **Industrial, Commercial, and Agricultural** | <ul><li>Opportunity costs of idle resources such as labor, land, and capital</li><li>Shutdown and restart costs</li><li>Spoilage and damage</li><li>Health and safety effects</li></ul> | <ul><li>Cost on other firms that are supplied by impacted firm (multiplier effect)</li><li>Costs on consumers if impacted firm supplies a final good</li><li>Health and safety related externalities</li></ul> |
| **Infrastructure and Public Service** | <ul><li>Opportunity cost of idle resources</li><li>Spoilage and damage</li></ul> | <ul><li>Costs to public users of impacted services and institutions</li><li>Health and safety effects</li><li>Potential for social costs stemming from looting, vandalism</li></ul> |

Induced costs include effects on social externalities and economic activities. The social costs associated with climate impacts on electrical systems can be diverse and widespread. For example, power outages during heat waves can expose vulnerable populations to increased morbidity and mortality.[13] Induced economic activities include the goods and services produced by companies using the power supplied by a utility. When a company that uses a utility's power must shut down or delay production due to an outage, that company's customers are also affected. Interdependent sectors (e.g., transportation, healthcare, and water) represent major sources of induced economic costs.

Electric service disruptions also significantly affect the reliability of other parts of the energy sector. These losses fall under indirect costs to commercial users, but because outages caused by climate impacts can be widespread and affect large geographic areas at once, they are of special concern. Failure of electrical equipment (e.g., electrical lines, pumps) can shut down steam boilers, cooling towers, pumps, and electrically operated safety control mechanisms in oil and gas refineries, pumping stations, terminals, and other facilities. Besides the lost revenue and other costs associated with equipment damage in these sectors, disruptions can lead to disruption in fuel deliveries (induced costs), worsening the effects of power outages for consumers. For example, following Hurricane Sandy in 2012, power outages caused widespread gasoline shortages in New Jersey and New York, limiting the ability of consumers to run generators.[14]

## SCALING CONSIDERATIONS: QUANTIFYING COSTS OF WIDESPREAD IMPACTS

Estimating the costs of climate and extreme weather events that cause widespread impacts requires careful consideration of regional variations in land, labor, and capital costs. These regional differences will affect both direct and indirect costs.

Land costs vary widely by region but may also fluctuate substantially within a region. Local variations in land costs can make accurate estimation of direct relocation costs extremely challenging. Regional variations in capital costs similarly affect the estimation of direct costs. Regional variations in labor rates affect estimates of labor costs for the direct repair and restoration of assets; more importantly, indirect cost estimates primarily predicated on the time value of money will vary substantially.

## 4.2 ANALYTICAL APPROACHES TO QUANTIFYING INDIRECT COSTS OF CLIMATE CHANGE IMPACTS TO RATEPAYERS

Quantitative estimates of indirect costs to utility customers as a result of climate change impacts are called Value of Lost Load (VOLL) calculations. Utilities can choose from a variety of approaches to determine VOLL, recognizing that estimates must be context specific and will vary by customer type. VOLL represents the value that customers place on reliable electricity service; it is also sometimes referred to as the Customer Damage Function (CDF) or the Value of Service Reliability (VOS).[15,16,17]

VOLL is usually measured in dollars per unit of power (e.g., megawatt hour, "MWh"). The VOLL depends on multiple factors, such as the type of customer affected, regional economic conditions and demographics, time and duration of outage, and other specific traits of the outage. As a result, while a rough "average" VOLL for a region can be estimated by analyzing available macroeconomic and electricity consumption data, an accurate estimate of VOLL requires surveying end-use customers in the region to determine their willingness to pay to avoid a specific type of outage.[18] Several methodologies for calculating VOLL are highlighted in Table 9.

Table 9. A summary of methods to calculate VOLL. [19,20,21]

| Approach | Description | Application |
|---|---|---|
| **Proxy Methods** | Uses observable variables linked indirectly to power supply security:<br>• Expenditure on standby generating facilities<br>• Monetized value of lost income and production output<br>• Other losses | • Suitable for cases in which anticipated losses can be expressed with sufficient precision using observed variables |
| **Case Studies/ Historical Data** | Performed after massive and major blackouts that affect large areas and large populations, causing serious and severe economic losses | • Yields most accurate and reliable data since these studies are conducted immediately after actual outage events.<br>• Rare and limited by geographic constraints as well as by the characteristics and duration of the outage; expensive strategy. |
| **Indirect Analytical Methods** **(*Macroeconomic Analysis*)** | Uses publicly declared and available, easy-to-reach and objective data to study outage costs. These data include GDP, annual energy consumption, peak power, and electricity tariffs. | • Easy, simple method; cheaper, less time-consuming, and highly objective<br>• Yields coarse results since all customer segments with distinct electric power consumption characteristics are analyzed together |
| **Customer Surveys** | After defining hypothetical outage scenarios and carefully designing a questionnaire, utilities ask customers to estimate the economic losses incurred during the predefined scenarios. | • Most popular tools chosen and utilized by the electric power industry and utilities to estimate outage costs |

In general, two types of VOLL can be estimated: marginal VOLL and average VOLL. Marginal VOLL measures the marginal value of the next unit of unserved power at peak periods (i.e., when customers place the highest value on power). Average VOLL represents the VOLL over a given period (e.g., month or year). Average VOLL tends to be lower than marginal VOLL, as it averages out the value that customers place on electricity over periods that include times when customers are not at home or businesses are closed. Average VOLL is commonly used to inform transmission and generation investment, where it may be more appropriate to estimate customers' willingness to pay over longer periods of time.

Lawrence Berkeley National Laboratory (LBNL) analyzed VOLL for different classes of customers across the United States in 2009 and updated the study in 2015. [22] The review draws on a variety of studies from different regions of the United States, including VOLL analyses following a survey approach. The review estimates interruption costs for different types of customers and for different outage durations, finding that costs increase as outage duration increases. However, maximum outage time reported is 16 hours, which may not capture costs associated with major outages, such as those that might follow an extreme storm surge event as part of a major hurricane. The VOLL estimates produced by this study are summarized in Table 10.

**Table 10.** Estimated interruption cost per event, average kilowatt (kW), and unserved kilowatt-hours (kWh; 2013 Dollars) by duration and customer class.[23]

| Interruption Cost | Interruption Duration | | | | | |
|---|---|---|---|---|---|---|
| | Momentary | 30 Minutes | 1 Hour | 4 Hours | 8 Hours | 16 Hours |
| **Medium and Large C&I (Over 50,000 Annual kWh)** | | | | | | |
| Cost per Event | $12,952 | $15,241 | $17,804 | $39,458 | $84,083 | $165,482 |
| Cost per Average kW | $15.9 | $18.7 | $21.8 | $48.4 | $103.2 | $203.0 |
| Cost per Unserved kWh | $190.7 | $37.4 | $21.8 | $12.1 | $12.9 | $12.7 |
| **Small C&I (Under 50,000 Annual kWh)** | | | | | | |
| Cost per Event | $412 | $520 | $647 | $1,880 | $4,690 | $9,055 |
| Cost per Average kW | $187.9 | $237.0 | $295.0 | $857.1 | $2,138.1 | $4,128.3 |
| Cost per Unserved kWh | $2,254.6 | $474.1 | $295.0 | $214.3 | $267.3 | $258.0 |
| **Residential** | | | | | | |
| Cost per Event | $3.9 | $4.5 | $5.1 | $9.5 | $17.2 | $32.4 |
| Cost per Average kW | $2.6 | $2.9 | $3.3 | $6.2 | $11.3 | $21.2 |
| Cost per Unserved kWh | $30.9 | $5.9 | $3.3 | $1.6 | $1.4 | $1.3 |

**Case Study: PG&E Service Interruption Costs in San Francisco Bay Area Storm Study**

In 2015, the Bay Area Council Economic Institute conducted a study to examine the region's vulnerability to a climate change-enhanced flooding event caused by an "atmospheric river" superstorm, Surviving the Storm.[24] In cooperation with the Bay Area Council and other project partners, the region's primary electric utility Pacific Gas and Electric Company (PG&E) estimated the value of lost service for a scenario in which six of the region's substations were disrupted during a flooding event. PG&E estimated that the indirect costs incurred by commercial customers (to temporarily relocate or continue their business operations) and residential customers (inconvenience) could total up to $125 million. The study noted that the impact would be mitigated by PG&E's redundant electric system where substations are interconnected through the electric grid and can support one another in order to help minimize customer service interruptions. Figure 10 shows the area flooded in the storm scenario, as well as the affected PG&E substations.

Figure 10. *Surviving the Storm scenario flooding and affected PG&E substations.*[25]

PG&E's indirect cost estimates did not include damage or spoilage costs or other induced costs. The cost estimate also does not include direct costs to PG&E, despite the assumptions about disruption to PG&E substations. The cost estimate aided PG&E and the Bay Area Council in creating an understanding of the scale of electricity outage costs relative to other storm-associated costs.[26]

## INTERRUPTION COST ESTIMATE CALCULATOR

LBNL developed an econometric model that can calculate customer interruption costs by season, time of day, day of the week, geographical region within the United States, and customer class. This Interruption Cost Estimation (ICE) Calculator is part of a publicly available tool[h] that uses ICE model results to calculate Value of Service Reliability (VOS), which is substantially equivalent to VOLL.

Use of the ICE Calculator has been described in case studies to demonstrate how the approach can support estimates of service reliability improvement value:

> **Electric Power Board (EPB) Chattanooga:** Using funding from a DOE Smart Grid Investment Grant, EPB Chattanooga deployed 1,200 automated circuit switches and sensors on 171 circuits to improve reliability across its entire service territory of about 174,000 homes and businesses. At a total cost of about $48.4 million, EPB substantially improved its reliability, reducing SAIDI[i] by 45% (from 112 to 61.8 minutes per

---

[h] http://www.icecalculator.com/

[i] System Average Interruption Duration Index. Equal to the sum of all customer interruption durations divided by the total number of customers served.

year) and reducing SAIFI[j] by 51% (from 1.42 to 0.69 interruptions per year). The ICE Calculator estimated the benefits of these improvements to consumers at about $26.8 million annually, in the form of avoided customer interruption costs.[27]

**Central Main Power (CMP):** In its 2014 rate case, CMP proposed to automate substations and line reclosers across its entire service territory (500,000 customers in southwest Maine), improving reliability with a 15-minute reduction in CAIDI (from 2.00 to 1.96 hours). Using the ICE Calculator, CMP calculated that the first six years of automation investments would deliver to CMP customers a Net Present Value of $20.7 million in avoided outage costs, more than double the NPV of the investment ($10.1 million).[28]

## CHAPTER 4 REFERENCES

[1] Ofori-Atta, K., Roseman, E., Saha, B., Stuart, S., Lipschultz, M., and Jonathan Smidt. 2004. "Profiting from Transmission Investment." *Public Utilities Fortnightly*. October. http://www.geni.org/globalenergy/library/technical-articles/finance/public-utilities-fortnightly/profiting-from-transmission-investment/PROFITING%20FROM%20TRANSMISSION%20INVESTMENT%20-%20OCT%2004%20-%20PUF.pdf.

[2] EEI (Edison Electric Institute). 2014. *Before And After The Storm. A compilation of recent studies, programs, and policies related to storm hardening and resiliency.* http://www.eei.org/issuesandpolicy/electricreliability/mutualassistance/Documents/BeforeandAftertheStorm.pdf.

[3] GAO (Government Accounting Office). 2014. *Energy Infrastructure Risk and Adaptation Efforts.* Washington DC: GAO. http://www.gao.gov/assets/670/660558.pdf.

[4] Lawton, L., M. Sullivan, K. Van Liere, A. Katz, and J. Eto. 2003. *A Framework and Review of Customer Outage Costs: Integration and Analysis of Electric Utility Cost Surveys.* Berkeley, CA: Lawerence Berkeley National Laboratory. November. https://emp.lbl.gov/sites/all/files/lbnl-54365.pdf.

[5] Markey, E., and H. Waxman. 2013. *Electric Grid Vulnerability- Industry Responses Reveal Security Gaps.* Washington, DC: Staff of Congressmen Edward Markey and Henry Waxman. May. https://portal.mmowgli.nps.edu/c/document_library/get_file?uuid=b2f47e65-330e-4d89-adee-e2ed58908927&groupId=10156.

[6] Feigel, R. E. 2013. "Cost Benefit Analysis of Power Reliability Strategies." *The Locomotive.* Hartford Steam Boiler Inspection and Insurance Company. https://www.hsb.com/TheLocomotive/CostBenefitAnalysisOfPowerReliabilityStrategies.aspx.

[7] Goulder L. H., and R. C. Williams. 2012. *The Choice of Discount Rate for Climate Change Policy Evaluation.* Resources for the Future. September. http://www.rff.org/files/sharepoint/WorkImages/Download/RFF-DP-12-43.pdf.

[8] Ofori-Atta, K., Roseman, E., Saha, B., Stuart, S., Lipschultz, M., and Jonathan Smidt. 2004. "Profiting from Transmission Investment." *Public Utilities Fortnightly*. October. http://www.geni.org/globalenergy/library/technical-articles/finance/public-

[j] System Average Interruption Frequency Index. Equal to the total number of customer interruptions divided by the total number of customers served.

utilities-fortnightly/profiting-from-transmission-investment/PROFITING%20FROM%20TRANSMISSION%20INVESTMENT%20-%20OCT%2004%20-%20PUF.pdf.

[9] Entergy. 2007. *Entergy Hurricane Hardening Study*. New Orleans, LA: Entergy, Inc. December.

[10] Entergy. 2007. *Entergy Hurricane Hardening Study*. New Orleans, LA: Entergy, Inc. December.

[11] Entergy. 2007. *Entergy Hurricane Hardening Study*. New Orleans, LA: Entergy, Inc. December.

[12] Adapted from Centolella, P. 2010. *Estimates of the Value of Uninterrupted Service for the Mid-West Independent System Operator*. SAIC. August. www.hks.harvard.edu/hepg/Papers/2010/VOLL%20Final%20Report%20to%20MISO%20042806.pdf.

[13] CDC (Centers for Disease Control and Prevention). 2013. "Heat-Related Deaths After an Extreme Heat Event — Four States, 2012, and United States, 1999–2009." *Morbidity and Mortality Weekly Report (MMWR)* 62(22): 433–436. Atlanta, GA: CDC. June. www.cdc.gov/mmwr/preview/mmwrhtml/mm6222a1.htm.

[14] DOE (U.S. Department of Energy). 2015. *Climate Change and the U.S. Energy Sector: Regional Vulnerabilities and Resilience Solutions*. Washington, DC: DOE. October. http://energy.gov/sites/prod/files/2015/10/f27/Regional_Climate_Vulnerabilities_and_Resilience_Solutions_0.pdf.

[15] Goel, L. and R. Billinton. 1994. "Prediction of Customer Load Point Service Reliability Worth Estimates in an Electric Power System." *IEE Proceedings on Generation, Transmission & Distribution* 141 (4): 390-396. July. http://ieeexplore.ieee.org/xpl/login.jsp?tp=&arnumber=296520&url=http%3A%2F%2Fieeexplore.ieee.org%2Fxpls%2Fabs_all.jsp%3Farnumber%3D296520.

[16] London Economics International LLC. 2013. *Estimating the Value of Lost Load*. Study for Electricity Reliability Council of Texas (ERCOT). Boston, MA: London Economics. June. www.ercot.com/content/gridinfo/resource/2015/mktanalysis/ERCOT_ValueofLostLoad_LiteratureReviewandMacroeconomic.pdf.

[17] LBNL (Lawerence Berkeley National Laboratory). 2015. *Updated Value of Service Reliability Estimates for Electric Utility Customers in the United States*. Berkely, CA: LBNL. January. https://emp.lbl.gov/sites/all/files/value-of-service-reliability-final.pdf.pdf.

[18] London Economics International LLC. 2013. *Estimating the Value of Lost Load*. Study for Electricity Reliability Council of Texas (ERCOT). Boston, MA: London Economics. June. www.ercot.com/content/gridinfo/resource/2015/mktanalysis/ERCOT_ValueofLostLoad_LiteratureReviewandMacroeconomic.pdf.

[19] Lawton, L., M. Sullivan, K. Van Liere, A. Katz, and J. Eto. 2003. *A Framework and Review of Customer Outage Costs: Integration and Analysis of Electric Utility Cost Surveys*. Berkeley, CA: Lawerence Berkeley National Laboratory. November. https://emp.lbl.gov/sites/all/files/lbnl-54365.pdf.

[20] London Economics International LLC. 2013. *Estimating the Value of Lost Load*. Study for Electricity Reliability Council of Texas (ERCOT). Boston, MA: London Economics. June. www.ercot.com/content/gridinfo/resource/2015/mktanalysis/ERCOT_ValueofLostLoad_LiteratureReviewandMacroeconomic.pdf.

[21] Kufeoglu, S., and M. Lehtonen 2015. "Interruption Costs of Service Sector Electricity Customers, a Hybrid Approach." *International Journal of Electrical Power & Energy Systems* 64: 588–595. January. www.sciencedirect.com/science/article/pii/S0142061514004815.

[22] LBNL (Lawerence Berkeley National Laboratory). 2015. *Updated Value of Service Reliability Estimates for Electric Utility Customers in the United States*. Berkely, CA: LBNL. January. https://emp.lbl.gov/sites/all/files/value-of-service-reliability-final.pdf.pdf.

[23] LBNL (Lawerence Berkeley National Laboratory). 2015. *Updated Value of Service Reliability Estimates for Electric Utility Customers in the United States*. Berkely, CA: LBNL. January. https://emp.lbl.gov/sites/all/files/value-of-service-reliability-final.pdf.pdf.

[24] BACEI (Bay Area Council Economic Institute). 2015. *Surviving the Storm*. San Francisco, CA: BACEI. March. http://documents.bayareacouncil.org/survivingthestorm.pdf.

[25] PG&E (Pacific Gas & Electric Company). 2016. *Climate Change Vulnerability Assessment*. San Francisco, CA: PG&E. January.

[26] PG&E (Pacific Gas & Electric Company). 2016. *Climate Change Vulnerability Assessment*. San Francisco, CA: PG&E. January.

[27] LBNL. 2015. *ICE Calculator Case Study Overview: EPB Chattanooga Distribution Automation*. Lawrence Berkeley National Laboratory.

[28] LBNL. 2015. *ICE Calculator Case Study Overview: CMP Distribution Automation*. Lawrence Berkeley National Laboratory.

# 5. ASSESS VULNERABILITIES

The final step in a vulnerability assessment requires a synthesis of the following three factors: the exposure of priority assets or operations to an adverse climate event (climate threat), the probability of damage to assets or disruption to operations exposed to those climate threats, and the likely consequences if the event were to occur (severity of impacts). Each of these factors is described in a preceding chapter:

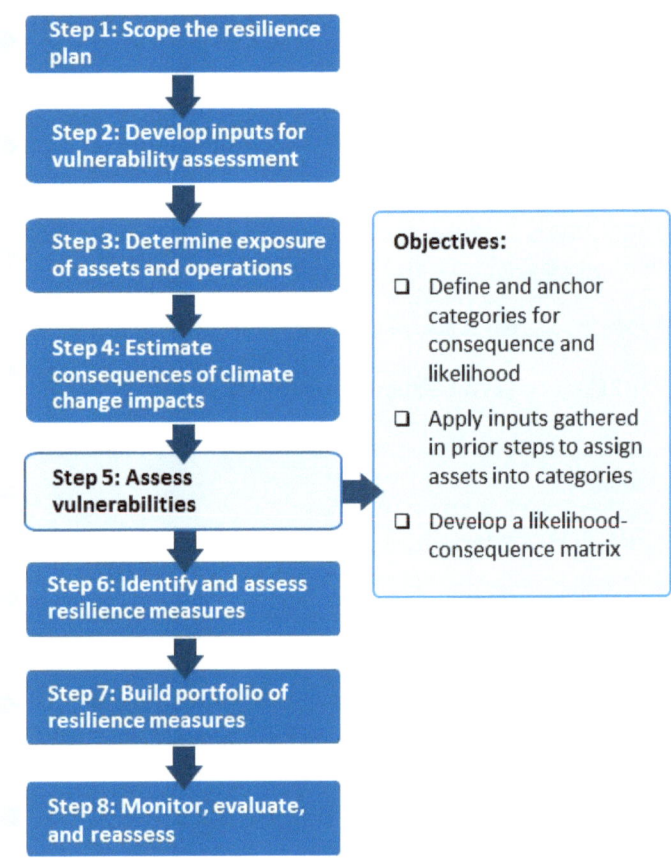

| | |
|---|---|
| Hazard/ Exposure (Ch. 3.1): | List of exposed assets and operations; brief description of climate threat |
| Likelihood/ Vulnerability (Ch. 3.2): | Annual probability of damage/disruption (e.g., Low, Medium, or High, with reasoning and confidence level) |
| Consequence /Cost (Ch. 4) | Severity of damage/disruption (e.g., Low, Medium, or High, with reasoning and confidence level) |

## 5.1 DEFINING AND DETERMINING RISK CATEGORIES

Several methods can be used to assess risks. In theory, a utility's risk profile is the sum total of risk from all individual events, and can be quantified according to Equation 1. In practice, utilities typically find it infeasible to develop a comprehensive set of event scenarios with quantitative estimates of likelihood and consequence—particularly given the current uncertainties associated with climate projections, limited actionable data, analytical challenges associated with processing and analyzing climate data, and other resource constraints. A more feasible approach may be to develop a likelihood-consequence matrix that uses qualitative categories of risk (e.g., high to low) to comparatively assess vulnerabilities and determine resilience priorities. The resulting matrix provides a straightforward guide for prioritization: risks with high consequence and low likelihood merit lower concern than those with high consequence and high likelihood.

(1) $f(x) = \sum_{i=1}^{\infty} (\text{Likelihood x Consequence})_i$

Qualitative categories make sense for several reasons. First, assigning the risk associated with climate change vulnerability is necessarily imprecise. Second, the level of analytical effort required to quantify both the likelihood and consequences of each climate-related threat may be prohibitive in some cases. Third, decision makers and stakeholders often find it easier to make decisions based on qualitative (high vs. low) factors than quantitative factors with statistical probabilities, particularly when there are high uncertainties with regard to likelihood or consequence.

Climate impact studies often use separate risk matrices to reflect a range of possible futures, such as "low emissions" and "high emissions" scenarios or multiple future scenarios (see Chapter 2). These ranges contribute to a better understanding of the probabilities associated with climate impacts on assets and operations. Generally, climate projections show greater agreement on the direction of change than on the magnitude or timeframe of change. For example, CMIP5 climate projections universally predict higher average temperatures for all regions in the United States, but they differ as to the extent and timing of temperature increases.[1]

Given the uncertainties inherent in projecting the precise timing and magnitude of future climate change, utilities may wish to consider all climate-related threats that could potentially affect their systems during the expected lifetime of an asset or capital investment. Based on the available information on consequences and likelihoods, utilities can sort their assets and operations into at least four main groups (more if additional qualitative categories are employed):

- **Low Likelihood/Low Consequence:** Assets/operations that have a low likelihood of being impacted by a future climate condition and, if impacted, would have a low consequence for system operations or performance.

- **Low Likelihood/High Consequence:** Assets/operations that have a low likelihood of being impacted by a future climate condition, but the impact would have a high consequence for system operations/performance.

- **High Likelihood/Low Consequence:** Assets/operations that have a high likelihood of being impacted by a future climate condition, but the impact would have a low consequence for system operations/performance.

- **High Likelihood/High Consequence:** Assets/operations that have a high likelihood of being impacted by a future climate condition and would have a high consequence for system operations or performance.

Utilities may decide to include more than two likelihood and consequence categories, particularly if the categories allow meaningful differentiation for ranking and prioritization purposes. Five or more categories provide better dispersion but may imply an inflated level of precision or may not provide sufficient added value to justify the extra effort that would be required. Every utility is unique, and the number of categories should reflect the available data and complexity of the company.

## Towards Quantitative Climate Risk Evaluation

Efforts are underway at DOE and elsewhere to develop information, methods, and tools for improving quantitative evaluation of climate vulnerabilities and resilience strategies. The first installment of DOE's Quadrennial Energy Review recognized the importance to address this issue and recommended several actions to address gaps, including:[2]

- **Develop comprehensive data, metrics, and an analytical framework for energy infrastructure resilience, reliability, and asset security:** Multiple gaps in federally accessible data impede decision-making on policies and investment related to resilience, reliability, and security. These data are critical for understanding the extent to which our existing energy infrastructure is resilient and for better informing resilience investments. DOE, in collaboration with DHS and interested infrastructure stakeholders, should develop common analytical frameworks, tools, and metrics to assess the resilience, reliability, and security of energy infrastructures. The purpose of this work will be to help inform, coordinate, set priorities for, and justify expenditures across federal agencies to increase the resilience, reliability, and security of energy infrastructure.

- **Value new services and technologies:** Efficient characterization and valuation of services provided to the grid by existing and new technologies is important for maintaining reliability and affordability of the rapidly evolving electricity system and providing clear price signals to consumers. Existing methods for establishing values and rates should appropriately compensate new technologies, with the potential to more effectively provide grid services reliably, affordably, and in compliance with environmental regulations. The Federal Government can play a role in developing frameworks to value grid services and approaches to incorporate value into grid operations and planning.

## RISK CATEGORY ANCHORS

When developing each risk category, establishing a clear descriptive parameter—or *anchor*—will foster consistent interpretation among individuals providing input to the assessment or viewing the results. The more descriptive the category anchors, the more consistent their application and interpretation will be. Chapter 6 explains how these categories are used. Understanding their usage can help determine the appropriate number of categories and a suitable level of detail for the anchors. The aim is to find the right balance between simplicity and detail. Table 11 provides example category descriptions.

**Table 11.** Illustrative anchors for likelihood and consequence parameters for hypotethical risk categories.

|  | Likelihood | Consequence |
|---|---|---|
| **High** | Once in 2 years or less | Cost of $100 million or more |
| **Medium** | Once in 2 to 25 years | Cost of $1 - 100 million |
| **Low** | Once in 25 years or more | Cost of less than $1 million |

## 5.2 ASSIGNING EXPOSED ASSETS/OPERATIONS INTO RISK CATEGORIES

Once the likelihood-consequence categories have been determined, exposed asset or operations should be assigned to the categories with the most closely aligned descriptive anchors, as discussed above. This process can be among the more labor-intensive steps in the vulnerability assessment, depending on the level of detail calculated in Chapters 3 and 4. Within each category, assets and operations can be ranked based on the exposure to climate threats and extreme weather, the estimated probability of a given climate event occurring, the estimated probability of damage or disruption, and the value of likely consequences.

If an asset or operation is exposed to multiple climate threats (as determined in Chapter 3), the asset/operation should be assigned into risk categories for each threat. For example, if a transmission line segment is exposed to threats from both wildfire and from more intense heat waves, then the transmission line segment should be placed into risk categories for both threat scenarios. This risk might be placed in low likelihood/high consequence risk categories for wildfire threat and placed in high likelihood/low consequence risk categories for the threat of more intense heat waves.

Common techniques for categorizing the likelihood and consequence include conducting individual analyses, holding workshops to collectively analyze risks, and soliciting judgments through structured surveys:

**Individual Analysis:** Categorizing each asset or operation vulnerable to climate and weather threats involves identifying in-house staff and others knowledgeable about the utility's operations and qualified to make judgments about the proper categorization of likelihoods and consequences. This method is often relatively fast but limits broader perspectives on any vulnerabilities that may be particularly complex or otherwise difficult to categorize (e.g., emerging climate-related threats). Individual analyses are most effective when the scope of risks is relatively specialized (not requiring cross-functional discussion) or when categorization is relatively intuitive (so that a single person could make adequate judgments).

**Workshops:** Workshops provide a collective technique for placing assets/operations into risk categories, bringing together a diverse set of informed perspectives. Staff and experts with a range of experiences and insights can collectively discuss and determine appropriate categorization. For example, a facilitated workshop might involve climate experts, facility managers, asset operators, and corporate risk management personnel to discuss and make judgments on suitable categories.

**Structured Surveys:** Surveys are another technique for assigning risk categories. Surveying knowledgeable staff and other experts can be useful for large, geographically distributed companies. Although surveys can be a useful alternative to workshops in some cases, the level of detail and quality of data can be less reliable if the surveys are not carefully designed and executed. For example, surveys do not easily allow respondents the opportunity to ask clarifying questions or engage in cross-functional discussion.

In many cases, a combination of these techniques can provide an initial categorization of likelihood and consequence. Some categorization efforts may be appropriate for individual analyses or surveys, while others may require group involvement via workshops.

When assigning categories, it is important to record key assumptions and characterize the associated level of confidence. Risks surrounded by greater uncertainty may prompt decision makers to ask for more information before investing resources to mitigate them. As discussed in Chapters 6 and 7, some risks should be reassessed as new information arises, often during the prioritization of mitigation activities.

## SYSTEM INTERDEPENDENCIES

Electric grid and energy system interdependencies may affect the likelihood and consequence of climate impacts on vulnerable assets and operations. A thorough assessment of risks should examine the effects on system performance from the loss or impairment of these assets due to climate change and extreme weather events. More broadly, the ability of a utility to continue operating in the event of extreme weather or changing climate depends on the abilities of its suppliers to continue supplying fuel, replacement parts, and other crucial equipment. A comprehensive vulnerability assessment will consider potential climate impacts on critical elements of the entire supply chain, including suppliers and customers (see 3.1, Supplier and Connected-Infrastructure Vulnerabilities).

## LIKELIHOOD–CONSEQUENCE MATRIX

Exposed assets/operations can be displayed in a two-dimensional matrix, with each quadrant reflecting a unique pairing of likelihood and consequence. Color-coding of the matrix helps in visualizing risks by category. If two categories (low and high) are used, the matrix might resemble Figure 11. If desired, the size or color of data points representing each asset or operation may be scaled to parameters such as the speed of onset or degree of uncertainty in the estimates. The matrix can be useful in providing decision makers with a visual perspective on the relative likelihood and consequence of each exposed asset/operation, and it may be useful to help screen which risks to prioritize.

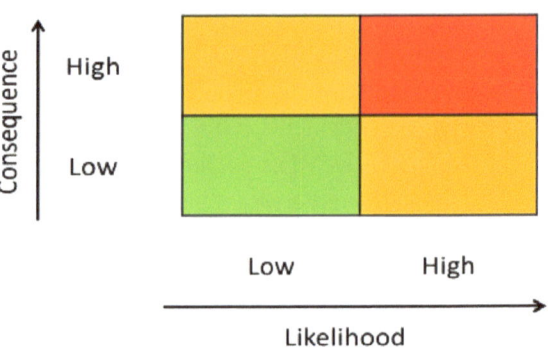

**Figure 11.** Sample likelihood–consequence matrix.

**Case Study: Northern Powergrid Risk Matrix**

Northern Powergrid developed a matrix to define relative climate risks using four consequence categories (negligible, marginal, critical, and catastrophic) and four likelihood categories (improbable, possible, probable, and near certain). Consequences were based on the anticipated business effects of a climate event damaging or disrupting an asset. Risks were assessed based on the combined effects of the likelihood and consequence of 13 different climate or extreme weather impacts (AR1- AR13, see Figure 12) such as transformers being derated due to high temperatures or overhead lines being affected by overgrown vegetation (due to a prolonged growing season). Mapping the likelihood of a climate event and the magnitude of resulting business consequences helped Northern Powergrid prioritize risks. Ultimately, this matrix was used to inform decisions about resilience measures.[3]

**Assessed Risks:**

**AR1:** Overhead line conductors affected by temperature rise, reducing rating and ground clearance.

**AR2:** Overhead line structures affected by summer drought and consequent ground movement.

**AR3:** Overhead lines affected by interference from vegetation due to prolonged growing season.

**AR4:** Underground cable systems affected by increase in ground temperature, reducing ratings.

**AR5:** Underground cable systems affected by summer drought and consequent ground movement, leading to mechanical damage.

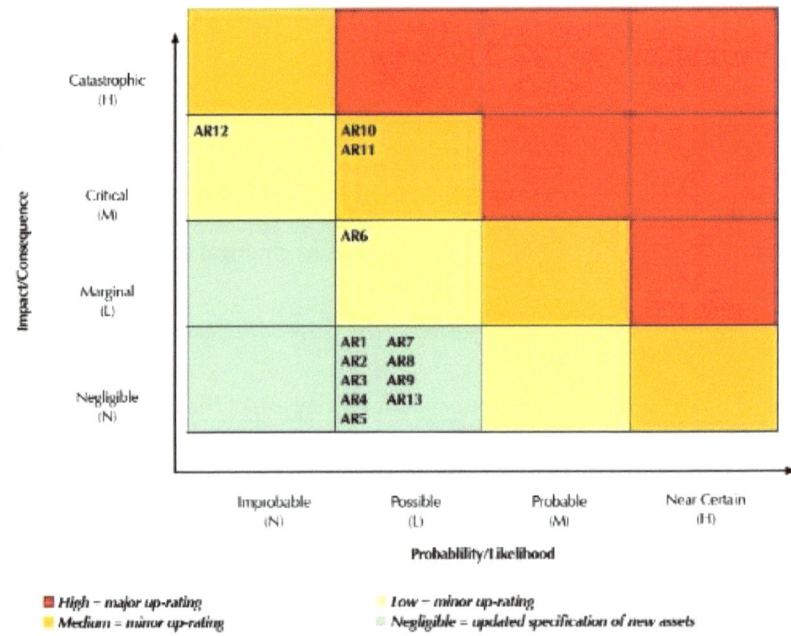

**Figure 12.** Risk Matrix created by Northern Powergrid to assess vulnerabilities of business operations to climate change.[4]

**AR6:** Substation and network earthing systems adversely affected by summer drought conditions, reducing the effectiveness of the earthing systems.

**AR7:** Transformers affected by temperature rise, reducing rating.

**AR8:** Transformers affected by urban heat islands and coincident air conditioning demand, leading to overloading in summer months.

**AR9:** Switchgear affected by temperature rise, reducing rating.

**AR10:** Substations affected by river flooding due to increased winter rainfall.

**AR11:** Substations affected by pluvial (flash) flooding due to increased rainstorms in summer and winter.

**AR12:** Substations affected by sea flooding due to increased sea levels and/or tidal surges.

**AR13:** Overhead lines and transformers affected by increased lightning activity.

## CHAPTER 5 REFERENCES

[1] CMIP5 (Coupled Model Intercomparison Project). 2016. Program for Climate Model Diagnosis and Intercomparison. Lawrence Livermore National Laboratory. http://cmip-pcmdi.llnl.gov/cmip5.

[2] DOE (U.S. Department of Energy). 2015. *Quadriennial Energy Review: First Installment*. Washington, DC: DOE. April. http://energy.gov/epsa/downloads/quadrennial-energy-review-first-installment.

[3] Northern Powergrid. 2012. *Climate Change Adaptation Report*. Northern Powergrid. http://www.northernpowergrid.com/asset/0/document/194.pdf.

[4] Northern Powergrid. 2012. *Climate Change Adaptation Report*. Northern Powergrid. http://www.northernpowergrid.com/asset/0/document/194.pdf.

Climate resilience planning is a three-part process. The first part, the establishment of goals, is a critical starting point that involves establishing the scope of the process. The second part, the vulnerability assessment, involves determining where the system is vulnerable and under what conditions. The third part of the process, identifying resilience solutions and developing the resilience plan, begins with Chapter 6. The resilience plan relies on information generated or assembled during the vulnerability assessment (described in Chapters 2–5), including the probabilities of adverse climate events, threshold conditions likely to affect important assets or overall system performance, and the consequences or costs of climate impacts.

The resilience plan prioritizes a set of actions or resilience measures to mitigate critical vulnerabilities. A range of resilience measures may be available for each asset or vulnerability to either reduce the

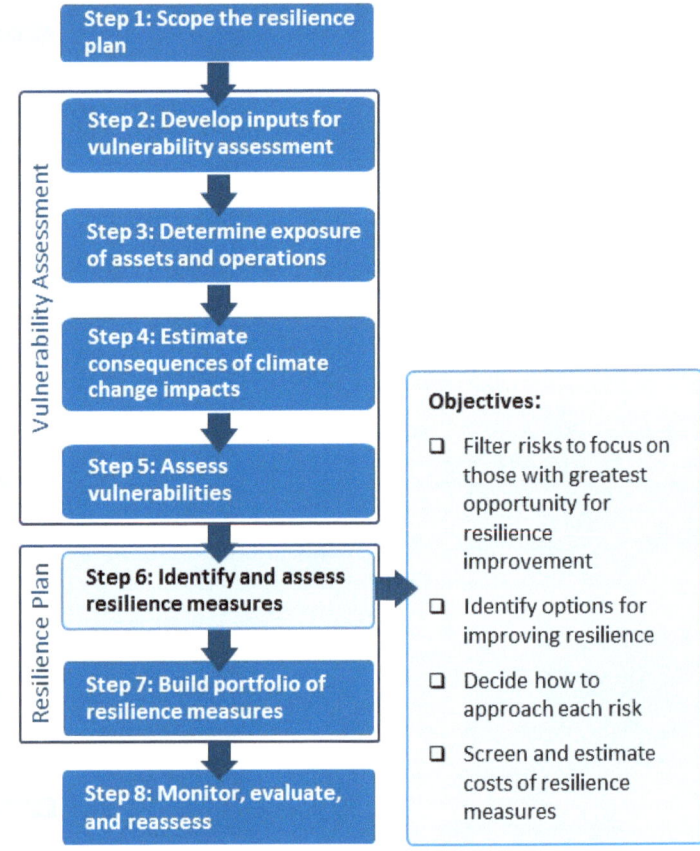

**Objectives:**

☐ Filter risks to focus on those with greatest opportunity for resilience improvement

☐ Identify options for improving resilience

☐ Decide how to approach each risk

☐ Screen and estimate costs of resilience measures

probability of damage and disruption (e.g., hardening and relocating assets) or reduce the consequences of any damage or disruption (e.g., recoverability and risk transfer/insurance). This chapter discusses the selection of measures to include in the resilience plan. Specifically, it provides guidance on identifying and examining the range of resilience options and determining the costs and impacts of each.

The costs of resilience measures may be substantial. The high cost of some resilience measures and uncertainties regarding risk complicate investment choices and highlight the importance of logically and systematically determining the costs and benefits of resilience solutions—and of business as usual. While the costs of climate resilience actions may be significant, the costs of inaction may be even greater.

## APPROACHES TO INVESTIGATING RESILIENCE MEASURES

Following the methodology outlined in the previous chapters, utilities will have identified their main vulnerabilities and characterized their consequence and likelihood. In terms of prioritizing resilience actions to pursue, one approach is to conduct a preliminary investigation into resilience measures to gain some familiarity with the options available. Knowing what options exist can help planners determine whether mitigating the risk is likely to be worth the investment. Based on the preliminary investigation, utilities can screen out risks for which the available resilience measures seem unlikely to provide benefits that outweigh costs. This approach generates a reduced set of risks and may facilitate identification of appropriate resilience actions that yield benefits outweighing costs. Utilities should filter the set of risks to focus resources where they can deliver the greatest improvement in resilience.

Using this approach, utilities would first identify possible resilience measures and take a preliminary look at the costs and benefits as well as the political, operational, and technical feasibility of each potential resilience measure. Actions deemed infeasible or not worth the investment could then be eliminated from consideration. The remaining potential measures could be evaluated with more detail and analysis of cost and benefit information.

## 6.1 DETERMINE POTENTIAL RESILIENCE MEASURES

A wide range of measures can improve the resilience of electricity assets and systems to climate change impacts. These measures include making physical and structural improvements to "harden" the system components as well as planning and modifying operations to build resilience.

> **System Hardening:** Elevate, retrofit, or reinforce existing structures; relocate assets; restore coastal wetlands or other natural barriers; enable greater distributed generation, islanding, and microgrids

> **Planning and Modifying Operations**: Update designs and resource plans; implement smart grid communications and monitoring technologies that improve grid observability and controllability; implement energy efficiency programs; enhance vegetation management; inventory spare parts; deploy demand response management tools; engage in mutual aid agreements; purchase risk transfer/insurance

Resilience measures may be generalized during a preliminary investigation stage, but utilities will need to consider the specific site characteristics of individual assets and systems when it is time for detailed analysis. The main types of resilience measures are described briefly below and in Appendix B.

## HARDENING EXISTING ASSETS

Hardening measures include initiatives to make physical and structural improvements to lines, poles, towers, substations, generation and supporting facilities, including elevating existing equipment or building and reinforcing floodwalls. There are a number of examples of hardening involving the application of design standards, construction guidelines, maintenance routines, inspection procedures, and adoption of innovative technologies.[1]

> **Targeted undergrounding**: Utilities may selectively underground lines to reduce exposure to lightning, tree and storm damage, and doing so by evaluating targeted undergrounding opportunities to maximize the benefit, given the added costs of undergrounding.

> **Strengthening transmission and distribution lines:** As an alternative to undergrounding, overhead lines can be strengthened by adding structural reinforcement (e.g., steel poles, guy wires, pole treatment) to existing lines. In addition, breakaway cables can be installed to avoid cascading pole system failures and minimize the restoration effort.

> **Hydrophobic coatings:** Special hydrophobic coatings help reduce damage to transmission and distribution system components by shedding water and facilitating ice removal. These coatings are already being used in some applications.

**Floodwalls and elevating key assets**: Utilities can reduce vulnerabilities to sea level rise, storm surge and floods by elevating existing and new equipment, building floodwalls to prevent exposure, and increasing the use of submersible equipment (e.g., substations, transformers, switches, pumps, etc.). Hardening against flooding and inundation can also include sealing conduits and cable penetrations, and shrink-wrapping cabinets and weatherproofing enclosures.

**Advanced water cooling technologies for thermoelectric generation**: Power plants require significant volumes of water for thermoelectric cooling. Utilities can employ alternative approaches to once-though cooling technologies to reduce their water use, including recirculating cooling, dry cooling, and wet-dry hybrid cooling technologies.

Measures that limit the number of customers affected by outages can also "harden" the grid. Examples include installing additional substations, as well as expanded use of distributed generation, microgrids capable of islanding, and load management programs. Illustrative examples include:

**Distributed generation**: Increased use of distributed generation (e.g., PV solar, wind, fuel cells, plug-in electric vehicles, etc.) can provide additional capacity to enhance resilience particular during periods of major outages. In some cases, these systems can disconnect from the bulk power system and serve as an independent backup power system.

**Microgrids**: Microgrids consisting of distributed generation, storage and energy management and control systems can be configured to operate in unison with the bulk power grid during most times, but operate independently as a complete, "islanded" electricity grid during outages.

**Remote monitoring and control**: Utilities can combine advances in automated monitoring and information technology to limit the number of customers affected by outages. Technologies such as reclosers, switches, and sectionalizers, limit the spread of outages and allow faster restoration of service to the unaffected sections of the lines.

Not all assets will be hardened or upgraded in the same way, as some resilience measures will be more cost-effective than others. For example, design and construction standards for upgrading or retrofitting existing assets are based on the local conditions of the facilities, so costs may vary regionally.[2]

Building protective features or relocating exposed assets to locations that reduce exposure to climate hazards can improve resilience. For a preliminary investigation of risks, a screening analysis of vulnerable sites or a record of repeated past impacts at a site may provide sufficient justification to consider hardening. Robust investigations would involve a detailed analysis of projected impacts for the location.

**Case Study: New York Power Authority (NYPA): Strategic Vision 2014–2019**

NYPA is planning a new system with traditional elements and innovative features like microgrids, clean distributed power sources near customer locations, and sophisticated smart grid devices. The goal is to improve reliability, resilience, and environmental protection and allow customers to manage their own power use. NYPA's Strategic Vision is developed around three key themes that reflect the many changes in the energy industry and the economy: customer empowerment, infrastructure modernization, and resource alignment. The plan also specifies steps NYPA will take in the short-, medium-, and long-term to incorporate climate resilience measures. [3]

## PLANNING AND OPERATIONS

Given the long service lifetimes of most energy infrastructure, energy sector managers and investors are experienced at planning in the presence of risk and uncertainty. Recognizing these long planning horizons, resilience planning should seek to extend system flexibility such that systems are able to handle a range of possible future conditions. Examples of planning and operational measures to improve resilience include risk mitigation actions such as upgrading communications equipment, managing vegetation, acquiring backup generators and other standby equipment, and improving or creating new emergency operations plans and mutual assistance groups. [4]

**Siting and design standards:** Design standards for new lines, poles, substations, and other transmission and distribution equipment can improve resilience over the long term at a much lower cost than expensive retrofits. Siting power lines to avoid high-risk areas and choosing designs and configurations that are resilient to flooding, fire, or wind will help avoid future disruptions.

**Vegetation management:** Modification of vegetation management programs to increase the frequency and extent of trimming can be an effective means of reducing line strikes. Some utilities are undertaking additional clearing vegetation on their easements and working with adjoining property owners to remove additional vegetation based on information collected from past storm damage.

**Load management:** Load reduction measures can help reduce outages and aid restoration, and can be achieved through a number of approaches including voluntary load-reduction programs, direct load control, and time-of-use tariffs.

**Damage prediction and response:** Advanced weather models can be used to predict when and where disruptions or damage may occur. Utilities can conduct studies of climate- and weather-related outages to better understand how wind, precipitation, and other important meteorological parameters are related to past system failures, and use these models to pre-position physical and human assets.

**Restoration management:** Like damage prediction, procedures and systems that allow utilities to shift from centralized to decentralized restoration management can improve response and restoration times.

Many utilities carry out a medium- to long-range strategic planning process, during which they discuss general changes in the electric industry, environmental pressures, technology changes, and the way asset owners plan to accommodate these changes over the next 5 to 20 years. [5] One type of process used by utilities is called an integrated resource plan (IRP). Typically the main objective of an IRP is to ensure projected electricity demand (plus a reserve margin) will be met over a set period of time. Some of the elements of IRPs—such as load forecasts, reliability, and supply options—may be affected by changes in climate and extreme weather events. These planning processes provide an opportunity for utilities to change their planning and asset management to build climate resilience. To minimize costs and distribute the timing of improvement projects, many utilities typically implement resilience investments in their plans as part of routine infrastructure improvement efforts (e.g., selecting less vulnerable locations or more-resilient components during scheduled replacement or maintenance of energy infrastructure).

**Case study: PSEG Emergency Restoration Policy**

PSEG Long Island's Emergency Restoration Implementation Procedures and Logistics Support Emergency Procedures dictate the utility's response to large-scale storms and other disasters involving equipment failure. Procedures include storm anticipation actions (e.g., placing remaining segments of the barrier containment system for flood control at the substations that may experience flooding, preparing all substations for storm conditions by securing loose items, removing any scaffolding, and tying down material and equipment), crew guide instructions, and actions specified in the Mutual Assistance Crew's Guidebook.[6]

**Case Study: An Integrated Resource Plan for the Entergy Utility System and the Entergy Operating Companies, 2009–2028**

Entergy sought to upgrade its generation and power supply resources to provide a more diverse, modern, and efficient portfolio of energy sources to meet customer needs. The company developed a strategic resource plan (SRP), which includes a set of principles and objectives that guide long-term portfolio design. The SRP planning process created scenarios for potential future portfolio resource decisions, including resource timing, location, and technology. This plan describes key uncertainties for resource planning, such as power plant construction cost, environmental concerns, and market conditions. It emphasizes that decisions for actual resource development will be made as the plan is implemented over time and will depend on a range of factors affected by long-term uncertainty. An action plan is to be undertaken over the next one to five years.[7]

## ADDITIONAL RESOURCES

As more companies, institutions, and local and state governments engage in resilience planning, new information and best practices continue to be developed, including updates to resilience methods, technologies, and planning approaches. As a result, a growing collection of resilience planning resources is being made available, and several efforts to centralize and categorize these resources are listed here.

> **Adaptation Clearinghouse**: Developed by the Georgetown Climate Center, the Adaptation Clearinghouse seeks to assist policymakers, resource managers, academics, and others who are working to help communities adapt to climate change. The "energy resources" section makes available resources to help policymakers understand, plan, and prepare for impacts of climate change to the energy sector, ranging from changes in energy demand to preparing for threats to energy infrastructure. http://www.adaptationclearinghouse.org/sectors/energy/.

> **Climate Adaptation Knowledge Exchange (CAKE):** CAKE is a knowledge base of adaptation resources and resilience-building case studies from projects around the world managed by EcoAdapt. CAKE's resources can be sorted by sector, scale, and type of adaptation strategy. http://www.cakex.org/.

## 6.2 DETERMINING HOW TO APPROACH EACH RISK

Knowledge of potential resilience measures helps utilities decide how to approach each risk. The four general approaches are to mitigate, transfer, accept, or avoid risk (see box: Approaches for responding to risk). This guide focuses on risk mitigation—taking actions to reduce the likelihood and/or consequences of the climate threat. These risks are the ones that utilities cannot avoid or transfer (or those that utilities choose not to transfer) and do not want to accept.

Upon identifying potential resilience measures, a utility can decide how to proceed in mitigating priority risks. Some risks may warrant further investigation into appropriate resilience measures. In other cases, the identified measures may suggest no-regrets solutions—those that deliver such significant benefits that the utility should implement them regardless of the climate threat.

**Approaches for responding to risk**

Risk mitigation: Taking actions to reduce the likelihood and/or consequence of a risk.

Risk transfer: Shifting risk to another company or organization (e.g., buying an insurance policy). Risk transfer measures can be useful for addressing low-frequency, high-severity events; they can significantly reduce the risk reduction costs to mitigate rare events.

Risk avoidance: Shifting operations or goals so that the utility is no longer exposed to that risk (e.g., divesting assets particularly exposed to climate hazards). This approach is generally reserved for risks that are deemed major but for which reasonable resilience measures are not available and accepting or transferring the risk is not appropriate.

Risk acceptance: Operating as normal and dealing with impacts if/when they occur (i.e., business-as-usual). If the timeframe for expected impacts is beyond the utility planning horizon, the risk may be acceptable now but should be re-evaluated in the future. By default, risks not considered in the vulnerability assessment are "accepted," whether they are known or unknown.[8]

## 6.3 SCREENING RESILIENCE MEASURES

For risks that utilities seek to mitigate, a high-level screening of possible resilience measures can help focus further consideration. Depending on the type of asset or vulnerability, not all potential resilience measures may be applicable or effective. A screening process can effectively reduce the number of options through qualitative analysis, informed by stakeholder and expert input.

As with a vulnerabilities screening analysis, screening of resilience measures requires the identification of at least one parameter (or criterion) on which each measure or option will be evaluated. A first-order screening criterion is often based on the approximate costs to implement. Table 12 provides order-of-magnitude cost estimates for a sample of resilience options. These costs represent average ranges from available data; estimates vary substantially based on scale and location. Additional criteria to consider include benefits of the measure, political feasibility, the technological capability for implementation, and flexibility (i.e., the extent to which a measure can adapt to, or be revised or reversed in response to, changing conditions, needs, or regulatory requirements).[9,10] For example, some smart grid technologies might be screened out because of local regulatory conditions (e.g., jurisdictions with certain data privacy restrictions can limit deployment of some devices that speed identification of faults or enable islanding), or storm surge barriers may not be considered necessary or appropriate under local political conditions.

Once the set of potential resilience measures has been down-sized using appropriate screening criteria, utilities can consider the costs and benefits of the remaining options in more detail and broaden the focus to include other important criteria for a comprehensive evaluation of promising measures.

## SCALING CONSIDERATIONS

The costs of resilience measures are often affected by the specific attributes of a particular location or facility. This relationship may make the process of scaling up a screening analysis more complex and costly. In general, order-of-magnitude estimates may be sufficient for screening criteria, though the level of accuracy required will depend on the decisions to be informed, and cost estimates should be selected in consultation with appropriate stakeholders. For cost estimates with wide variation, utilities may need to conduct a series of analyses for similar regions to estimate the costs for the larger area. For example, the cost of undergrounding transmission lines may vary from $500,000 to $30,000,000 per mile, depending on utility- and location-specific factors.[11,12]

For detailed analyses, cost information may not be available for all locations, and new estimates would need to be calculated or collected. Public data often do not contain a specific breakdown of repair, relocation, and similar costs, so access to electricity asset-owner information can be valuable in developing accurate estimates.

## 6.4 DETERMINING COSTS OF RESILIENCE MEASURES

Utilities should focus the determination of costs on the screened list of potential resilience measures from the previous section. The focus should be placed on total costs, which include up-front capital costs as well as operating and maintenance costs over the lifetime of the resilience measure.

A summary of available cost information is provided below, including a range of costs for different example measures, largely drawing on DOE sources. Utility databases and experts will likely be able to provide additional detail on costs, especially those specific to local conditions.

## HARDENING EXISTING ASSETS

The costs of hardening existing assets and upgrades can span several orders of magnitude (see Table 12). While some of these measures are widely used by electric utilities, others are either new technology or not in common use and are therefore not widely discussed in the literature.

Relocation costs are primarily driven by real estate costs, type of construction required, and specific design parameters.[13] Estimates of real estate values for potential relocation sites may be obtained from local tax assessment records, while construction and design costs can be obtained from utility building departments or contracting firms.

Smart grid and microgrid capabilities may be among the more expensive resilience measures, with costs depending on the technology and project-specific context. These technologies are still developing, which means that much of the available initial investment and maintenance costs are not well documented.[14] While initial capital costs may be higher than some other resilience options, smart grid investments—like other options—may provide substantial co-benefits that should be considered (see Section 6.5). In the case of smart grid technologies, they can improve grid reliability even as systems are challenged by the two-way energy and intermittent flow from solar and wind generation and growing loads imposed by electric vehicle charging, while improving customer choice and reducing the environmental impact of electricity generation.

For ecosystem-based resilience measures, which include land restoration activities, integration of green infrastructure with engineered measures, and habitat protection, utilities may look into collaborating with managers or owners of local ecosystems to identify resilience measures and opportunities for cost sharing.

Table 12. Illustrative costs for selected resilience measures for utility assets.

| Example Resilience Measure | General Range or Example Cost | Notes/Sources |
|---|---|---|
| Guying | $600 to $900 per pole | [15] |
| Upgrade Wood Poles | $16,000 to $40,000 per mile | Depends on material (steel is more expensive than concrete); there are many possible upgrades in use (replace entire pole, replace wood cross-arms, reduce spans between poles).[16,17,18] |
| Submersible Equipment | >$130,000 per vault | Depends on location and type of submersible equipment needed.[19] |
| Upgrade Transmission Lines | >$400,000 per mile | Depends on specific upgrade.[20] |
| Substation Hardening | $600,000 per substation | Wide range of cost is available depending on specific hardening measure needed for each location.[21] |
| Elevating Substations | >$800,000 to >$5,000,000 to elevate | Difficult to determine due to variation in height needed for each location.[22,23] |
| Reinforce Floodwall | $220,000 per mile | Based on 36-mile Port Arthur seawall. Costs depend on site-specific factors such as material composition, thickness, height, geology, and location of floodwall.[24] |
| Build New Floodwalls | $4,000,000 per mile | Depends on site-specific factors as noted above.[25] |
| Undergrounding Distribution Lines | $100,000 to $5,000,000 per mile | Depends on area (urban is most expensive) and new construction or conversion from overhead (new construction is more expensive).[26,27,28] |
| Undergrounding Transmission Lines | >$500,000 to $30,000,000 per mile | Depends on area (urban is generally more expensive) and new construction or conversion from overhead (new construction is more expensive).[29,30,31] |
| Install Microgrid | $150,000,000 for 40MW average load | Depends on size of the microgrid and the average load needed; this is a not yet deployed widely so costs are uncertain.[32] |

| Example Resilience Measure | General Range or Example Cost | Notes/Sources |
|---|---|---|
| Advanced Metering Infrastructure | $240 to >$300 per smart meter installed | Depends on the size of the network and the number of meters installed; this is a new technology that is still developing, so costs are uncertain. [33] |
| Marsh Stabilization | $2 per square meter | [34] |
| Marsh Creation | $4.30 per square meter | [35] |

## PLANNING AND OPERATIONS

Planning and operations measures are often less expensive than many engineering-based resilience measures. In addition to the measures listed below in Table 13, other planning activities, such as long-range strategic planning, updating emergency operations plans, and participating in mutual assistance groups, can be important components of a more resilient utility. [36]

**Table 13.** Illustrative costs for selected resilience measures for utility operations.

| Example Resilience Measure | General Range or Example Cost | Notes/Sources |
|---|---|---|
| Vegetation Management | $12,000 per mile | Depends on the functionality of the existing vegetation management plan in place and the level of vegetation clearing that the utility chooses (tree maintenance, tree removal, enhanced tree trimming vs. routine tree trimming). [37,38,39] |
| Backup Generators | $20,000 per substation | Depends on the size of the substation and the amount of power needed in a backup situation. [40,41] |
| Demand Reduction Programs | $50 to >$1,000 per MWh | Includes appliance recycling programs, demonstrations, education initiatives, weatherization incentives, and similar consumer behavior programs. [42] |

**Case Study: Potomac Electric Power Company (Pepco) Reliability Investments**

In 2012, Pepco proposed specific reliability investments—including improvements to priority feeders, accelerating tree trimming, and undergrounding overhead distribution feeders. This plan was developed in response to the Report of the Grid Resiliency Task Force. [43] The Task Force Report contains eleven recommendations, including four for which it urged immediate action "to accelerate resiliency improvements and provide Marylanders with a tangible benefit in a short period of time." Provided below are examples of actions identified in the report to improve resilience.

**Key measures to improve reliability:**

1.  Improve priority feeders, which involved upgrading and hardening 24 overhead distribution feeders over two years to improve the performance as measured by SAIFI and SAIDI

2. Accelerate tree trimming
3. Underground overhead distribution feeders

Pepco proposed a "grid resiliency charge" to recover the costs of accelerated capital and operations and maintenance projects resulting from currently planned reliability work. This charge will be in effect only until the incremental project costs are incorporated into Pepco's base rates. Pepco proposed a customer credit if it does not meet the minimum reliability standards and an incentive for achieving the accelerated reliability standards.[44,45]

**Table 14.** Pepco Maryland grid resiliency work – estimated cost.[46,47]

| Project | Scope | Overall Cost | Specific Cost | Duration |
|---|---|---|---|---|
| **Priority Feeders** | Upgrading and hardening 24 distribution feeders | $12-million-per-year capital investment | $1 million per feeder | Two years (2014 and 2015) |
| **Vegetation Management** | Accelerating the four-year trim cycle of scheduled clearance tree trimming to three years | $17 million O&M expense | No details provided | One year (2014) |
| **Selective Undergrounding** | Undergrounding six 13 kV distribution feeders | $151 million capital investment | Estimated $25 million per feeder | Three years (2013–2016) |

## 6.5 ASSESS POTENTIAL BENEFITS OF RESILIENCE MEASURES

Resilience measures may provide a variety of benefits, including direct benefits from avoided costs (based on potential costs of impacts), as well as co-benefits (e.g., system reliability benefits, enhanced energy efficiency, reduced GHG emissions, etc.). Capturing the value of benefits is difficult. Utilities should consider economic and non-economic metrics appropriate for the decision context and requirements. Since the primary direct benefits of resilience measures are the avoided potential costs of climate impacts, which are discussed in Chapter 4, this section briefly summarizes the avoided costs and focuses on potential metrics and qualitative considerations for additional benefits, where available. A diverse set of metrics can help to inform the overall value (economic and non-economic) of investing in resilience measures.

### AVOIDED DIRECT AND INDIRECT COSTS OF IMPACTS

As discussed in Chapter 4, direct costs of climate impacts on electric utilities can be assessed by economic loss due to damage and disruption to assets and operations and the associated repair or replacement costs. Potential indirect costs can include customer losses associated with interrupted power, as well as any damaged customer equipment.

Resilience measures can provide benefits (avoid incurred costs) not only to particular assets but also to the broader electricity systems. Some of these benefits can be captured through reliability and resilience metrics. A variety of metrics exists to measure electricity system reliability at the distribution level, which generally apply to interruptions or outages of less than 24 hours. Further development is needed to understand applicability to potential outages of longer duration possible with very high-impact, low-frequency events. However, there is not a generally agreed-upon method to quantify the resilience of a system. A variety of resilience metrics can help to assess the resilience of electricity systems and provide insights into the system-level benefits of resilience measures. Most metrics are based on measuring reliability, which can be used as a proxy for some elements of resilience. Examples of reliability metrics for distribution systems include the following:

> **System Average Interruption Frequency Index (SAIFI):** A measure of the average frequency of interruptions per total number of customers. It is the number of interruptions divided by the total number of customers served.

> **System Average Interruption Duration Index (SAIDI):** A measure of the average duration of service interruptions for the total number of a utility's customers. It represents the minutes interrupted divided by total number of customers served.

> **Customer Average Interruption Duration Index (CAIDI):** The average outage duration that any given customer would experience. It represents the minutes interrupted divided by the number of customers affected. It can also be viewed as the average restoration time.

> **Customer Restoration-90 (CR-90):** The number of hours it takes from the start of the outage event to restore power to 90% of the affected customers of a given utility. This metric is designed specifically to apply to consideration of major high-impact events during which power is lost to a large number of electric customers.

The following resources provide additional information on these and other methods for measuring reliability and resilience:

> H.H. Willis and K. Loa. 2015. The RAND Corporation. *Measuring the Resilience of Energy Distribution Systems.* http://www.rand.org/pubs/research_reports/RR883.html.

> M. Keogh and C. Cody. 2013. *Resilience in Regulated Utilities.* The National Association of Regulatory Utility Commissioners (NARUC). http://www.ncsl.org/Portals/1/Documents/forum/Forum_2014/ResilienceRegulatedUtilities.pdf.

> U.S. Department of Homeland Security. 2013. National Infrastructure Protection Plan: Partnering for Critical Infrastructure Security and Resilience. https://www.dhs.gov/national-infrastructure-protection-plan.

> Watson et al. 2015. *Conceptual Framework for Developing Resilience Metrics for the Electricity, Oil and Gas Sectors in the United States.* http://energy.gov/sites/prod/files/2015/02/f20/EnergyResilienceRpt-Sandia-Sep2014.pdf.

**Case Study: Electric Power Board of Chattanooga**

Electric Power Board (EPB) of Chattanooga estimates benefits of about $26.8 million annually as a result of smart grid improvements, including installation of automated circuit switches and sensors (see also Section 4.2). EPB substantially improved its reliability, reducing SAIDI by 45% (from 112 to 61.8 minutes per year) and reducing SAIFI by 51% (from 1.42 to 0.69 interruptions per year).

During a single windstorm event in 2012, the utility estimates that automated fault isolation and service restoration technology the utility installed reduced the number of sustained outages by 50% to 40,000 customers. Reduced outages, as well as customer outage information provided by meters, helped the utility avoid 500 truck rolls and reduced total restoration time by 1.5 days.[48,49]

## CO-BENEFITS OF RESILIENCE MEASURES

In addition to avoiding costs from climate impacts and improving reliability, some resilience measures may provide co-benefits to other sectors, society, or ecosystems. Increased grid resilience can reduce expenditures by utilities and customers on items to mitigate the effects of power outages including back-up generators, second utility feeds, and power conditioning equipment. Similarly, some actions may be initially undertaken for an unrelated reason, but result in improved resilience for electricity infrastructure. In general, co-benefits to building resilience to climate change include improvements to economic growth and job creation, emergency management and preparedness, public health, national security, agricultural productivity, and ecosystem conservation.[50]

By expanding resilience plans to include resilience measures with possible co-benefits, utilities can lower the burden of resilience on strictly engineering and hardening investments. However, measures and data to determine the co-benefits of different actions have been very difficult to develop, especially for diffuse co-benefits to society.[51] When assessing benefits of resilience actions, utilities should consider—at least qualitatively—the potential co-benefits in evaluation of resilience measures.

Resilience measures with environmental co-benefits, such as wetlands restoration, may have low investment needs and high reduction potential of expected losses. Even if maintaining existing vegetation is not the most effective option in building resilience, positive co-benefits in other sectors could be a strong driver for implementation alongside more expensive measures.

It is increasingly recognized that many actions that enhance resilience to climate change and extreme weather can also contribute to reduced greenhouse emissions.[52] For example, measures that enhance energy efficiency and reduce energy demand improve resilience to increasing heat waves (which are likely to lead to higher air conditioning loads, higher peak demand, and higher losses on the transmission network) as well as reduce GHG emissions. Distributed generated clean energy sources also offer climate mitigation and resilience benefits. For example, solar PV and wind reduce the water intensity of energy generation (as compared to thermoelectric power generation), improving system resilience to reduced water availability and drought. Combined heat and power (CHP), which improves efficiency by using waste heat, can also improve resilience while reducing emissions. In addition, smart grid, microgrids, and distributed generation systems add resiliency within local distribution systems and may reduce the number of outages, the number of users affected by each outage, and the duration of outages. Locations with microgrids may also have key services up and running more quickly following an outage for the benefit of the overall community, including places of refuge. Table 15 provides several examples of climate resilience and climate mitigation co-benefits.

**Table 15.** Examples of resilience options with GHG mitigation co-benefits.

| Option/project | GHG Mitigation Benefit | Climate & Extreme Weather Resilience Benefit |
|---|---|---|
| **Distributed generation, including wind, solar PV, and CHP** | • Emits fewer GHG emissions than conventional fossil-based power sources | • Reduces customer dependence on broader electricity transmission and distribution grid<br>• Reduces dependence on generation sources that may be vulnerable to decreasing water availability |
| **Energy efficiency measures, including building codes** | • Reduces GHG emissions by decreasing demand for electricity generation | • Reduces potential for grid failure by decreasing energy demand during peak events (extreme heat or cold) |
| **Smart grids and microgrids** | • Can reduce GHG emissions by improving grid efficiency and enabling greater integration of renewable generation sources, energy storage, and electric vehicle charging | • Improves integration of renewable sources such as wind and solar PV, which are less vulnerable to decreasing water availability<br>• Reduces demand for long-distance transmission<br>• Improves quicker fault-locating and outage response times |
| **Energy storage** | • Reduces GHG emissions by enabling intermittent renewable sources such as solar PV and wind | • Improves ability to accommodate storm-related power outages and climate-related load peaks |
| **Green infrastructure, including cool roofs** | • Reduces GHG emissions by reducing electricity demand for cooling<br>• Reduces heat island effect | • Reduces potential for grid failure by decreasing energy demand during peak events (extreme heat) |

## CHAPTER 6 REFERENCES

[1] Electric Power Research Institute. 2013. Enhancing Distribution Resiliency: Opportunities for Applying Innovative Technologies. Palo Alto, CA: Electric Power Research Institute. January 11. http://www.epri.com/abstracts/Pages/ProductAbstract.aspx?ProductId=000000000001026889.

[2] EEI. 2014. *Before and After the Storm – Update: A Compilation of Recent Studies, Programs, and Policies Related to Storm Hardening and Resiliency*. Washington, DC: EEI. March. http://www.eei.org/issuesandpolicy/electricreliability/mutualassistance/Documents/BeforeandAftertheStorm.pdf.

[3] NYPA (New York Power Authority). 2014. *Strategic Vision 2014–2019*. NYPA. http://www.nypa.gov/PDFs/StraVis2014/C1B568998FA6919AE001FA29EBAAAD1F/STPLBK%209-236-13[1].pdf.

[4] Electric Power Research Institute. 2013. Enhancing Distribution Resiliency: Opportunities for Applying Innovative Technologies. Palo Alto, CA: Electric Power Research Institute. January 11. http://www.epri.com/abstracts/Pages/ProductAbstract.aspx?ProductId=000000000001026889.

[5] NYPA (New York Power Authority). 2014. *Strategic Vision 2014–2019*. NYPA. http://www.nypa.gov/PDFs/StraVis2014/C1B568998FA6919AE001FA29EBAAAD1F/STPLBK%209-236-13[1].pdf.

[6] PSE&G Long Island. 2014. *Electric Utility Emergency Plan.* Hicksville, NY: PSE&G Long Island. http://www.lipower.org/pdfs/stormcenter/2014EmergencyPlan.pdf.

[7] Entergy Corporation. 2010. *Building a Resilient Energy Gulf Coast: Executive Report.* www.entergy.com/content/our_community/environment/GulfCoastAdaptation/Building_a_Resilient_Gulf_Coast.pdf.

[8] EPA (U.S. Environmental Protection Agency). 2014. *Being Prepared for Climate Change: A Workbook for Developing Risk-Based Adaptation Plans.* Office of Water. Washington, DC: DOE. August. https://www.epa.gov/sites/production/files/2014-09/documents/being_prepared_workbook_508.pdf.

[9] DOE. 2016. *Climate Change and the Electricity Sector: Guide for Assessing Vulnerabilities and Developing Resilience Solutions to Sea Level Rise.* Draft, April.

[10] Economics of Climate Adaptation Working Group. 2009. *Shaping Climate Resilient Development: A Framework for Decision-Making.* Economics of Climate Adaption. http://media.swissre.com/documents/rethinking_shaping_climate_resilent_development_en.pdf.

[11] EEI. 2014. *Before and After the Storm – Update: A Compilation of Recent Studies, Programs, and Policies Related to Storm Hardening and Resiliency.* Washington, DC: EEI. March. http://www.eei.org/issuesandpolicy/electricreliability/mutualassistance/Documents/BeforeandAftertheStorm.pdf.

[12] TPUC (Texas Public Utilities Commission). 2009. *Cost-Benefit Analysis of the Deployment of Utility Infrastructure Upgrades and Storm Hardening Programs.* March. http://www.puc.texas.gov/industry/electric/reports/infra/Utlity_Infrastructure_Upgrades_rpt.pdf.

[13] DOE. 2016. *Climate Change and the Electricity Sector: Guide for Assessing Vulnerabilities and Developing Resilience Solutions to Sea Level Rise.* Draft, April.

[14] DOE (U.S. Department of Energy). 2010. *Hardening and Resiliency: U.S. Energy Industry Response to Recent Hurricane Seasons.* Washington D.C.: DOE. August. https://www.oe.netl.doe.gov/docs/HR-Report-final-081710.pdf.

[15] DOE (U.S. Department of Energy). 2010. *Hardening and Resiliency: U.S. Energy Industry Response to Recent Hurricane Seasons.* Washington D.C.: DOE. August. https://www.oe.netl.doe.gov/docs/HR-Report-final-081710.pdf.

[16] TPUC (Texas Public Utilities Commission). 2009. *Cost-Benefit Analysis of the Deployment of Utility Infrastructure Upgrades and Storm Hardening Programs.* March. http://www.puc.texas.gov/industry/electric/reports/infra/Utlity_Infrastructure_Upgrades_rpt.pdf.

[17] DOE (U.S. Department of Energy). 2010. *Hardening and Resiliency: U.S. Energy Industry Response to Recent Hurricane Seasons.* Washington D.C.: DOE. August. https://www.oe.netl.doe.gov/docs/HR-Report-final-081710.pdf.

[18] FPL (Florida Power & Light Company). 2013. *2013–2015 Electric Infrastructure Hardening Plan.* Filed with the Florida Public Service Commission in Docket No. 130132-EI. May. http://www.psc.state.fl.us/library/FILINGS/13/02408-13/02408-13.pdf.

[19] FPL (Florida Power & Light Company). 2013. *2013–2015 Electric Infrastructure Hardening Plan.* Filed with the Florida Public Service Commission in Docket No. 130132-EI. May. http://www.psc.state.fl.us/library/FILINGS/13/02408-13/02408-13.pdf.

[20] TPUC (Texas Public Utilities Commission). 2009. *Cost-Benefit Analysis of the Deployment of Utility Infrastructure Upgrades and Storm Hardening Programs.* March. http://www.puc.texas.gov/industry/electric/reports/infra/Utlity_Infrastructure_Upgrades_rpt.pdf.

[21] FPL (Florida Power & Light Company). 2013. *2013–2015 Electric Infrastructure Hardening Plan*. Filed with the Florida Public Service Commission in Docket No. 130132-EI. May. http://www.psc.state.fl.us/library/FILINGS/13/02408-13/02408-13.pdf.

[22] TPUC (Texas Public Utilities Commission). 2009. *Cost-Benefit Analysis of the Deployment of Utility Infrastructure Upgrades and Storm Hardening Programs.* March. http://www.puc.texas.gov/industry/electric/reports/infra/Utlity_Infrastructure_Upgrades_rpt.pdf.

[23] Johnson, T. 2013. "PSE&G wants $1.7 Billion to Keep Electricity Substations High and Dry." *NJ Spotlight.* October. http://www.njspotlight.com/stories/13/10/23/pse-g-wants-1-7b-to-keep-electricity-substations-high-and-dry/.

[24] DOE (U.S. Department of Energy). 2010. *Hardening and Resiliency: U.S. Energy Industry Response to Recent Hurricane Seasons.* Washington D.C.: DOE. August. https://www.oe.netl.doe.gov/docs/HR-Report-final-081710.pdf.

[25] DOE (U.S. Department of Energy). 2010. *Hardening and Resiliency: U.S. Energy Industry Response to Recent Hurricane Seasons.* Washington D.C.: DOE. August. https://www.oe.netl.doe.gov/docs/HR-Report-final-081710.pdf.

[26] EEI. 2014. *Before and After the Storm – Update: A Compilation of Recent Studies, Programs, and Policies Related to Storm Hardening and Resiliency*. Washington, DC: EEI. March. http://www.eei.org/issuesandpolicy/electricreliability/mutualassistance/Documents/BeforeandAftertheStorm.pdf.

[27] TPUC (Texas Public Utilities Commission). 2009. *Cost-Benefit Analysis of the Deployment of Utility Infrastructure Upgrades and Storm Hardening Programs.* March. http://www.puc.texas.gov/industry/electric/reports/infra/Utlity_Infrastructure_Upgrades_rpt.pdf.

[28] DOE (U.S. Department of Energy). 2010. *Hardening and Resiliency: U.S. Energy Industry Response to Recent Hurricane Seasons.* Washington D.C.: DOE. August. https://www.oe.netl.doe.gov/docs/HR-Report-final-081710.pdf.

[29] EEI. 2014. *Before and After the Storm – Update: A Compilation of Recent Studies, Programs, and Policies Related to Storm Hardening and Resiliency*. Washington, DC: EEI. March. http://www.eei.org/issuesandpolicy/electricreliability/mutualassistance/Documents/BeforeandAftertheStorm.pdf.

[30] TPUC (Texas Public Utilities Commission). 2009. *Cost-Benefit Analysis of the Deployment of Utility Infrastructure Upgrades and Storm Hardening Programs.* March. http://www.puc.texas.gov/industry/electric/reports/infra/Utlity_Infrastructure_Upgrades_rpt.pdf.

[31] EEI (Edison Electric Institute). 2013. *Out of Sight, Out of Mind 2012: An Updated Study on the Undergrounding of Overhead Power Lines*. Prepared by Hall Energy Consulting, Inc. Washington, DC: EEI. http://www.eei.org/issuesandpolicy/electricreliability/undergrounding/Documents/UndergroundReport.pdf.

[32] Dohn, R. 2011. *The Business Case for Microgrids*. White Paper. Siemens AG. http://w3.usa.siemens.com/smartgrid/us/en/microgrid/Documents/The%20business%20case%20for%20microgrids_Siemens%20white%20paper.pdf.

[33] CEC (California Energy Commission). 2007. *Value of Distribution Automation Applications*. CEC 500-2007-028. http://www.energy.ca.gov/2007publications/CEC-500-2007-028/CEC-500-2007-028.PDF.

[34] Jones, H.P., D.G. Hole, and E.S. Zavaleta. 2012. "Harnessing Nature to Help People Adapt to Climate Change." *Nature Climate Change (2)*; pp. 504–509; doi: 10.1038/NCLIMATE1463. http://www.bios.niu.edu/jones/lab/Jones_et_al.2012.pdf.

[35] Jones, H.P., D.G. Hole, and E.S. Zavaleta. 2012. "Harnessing Nature to Help People Adapt to Climate Change." *Nature Climate Change (2)*; pp. 504–509; doi: 10.1038/NCLIMATE1463. http://www.bios.niu.edu/jones/lab/Jones_et_al.2012.pdf.

[36] DOE (U.S. Department of Energy). 2010. *Hardening and Resiliency: U.S. Energy Industry Response to Recent Hurricane Seasons*. Washington D.C.: DOE. August. https://www.oe.netl.doe.gov/docs/HR-Report-final-081710.pdf.

[37] TPUC (Texas Public Utilities Commission). 2009. *Cost-Benefit Analysis of the Deployment of Utility Infrastructure Upgrades and Storm Hardening Programs*. March. http://www.puc.texas.gov/industry/electric/reports/infra/Utlity_Infrastructure_Upgrades_rpt.pdf.

[38] PEPCO (Potomac Electric Company). 2010. *Comprehensive Reliability Plan for District of Columbia*, District of Columbia Order No. 15568. PEPCO. September. http://www.pepco.com/uploadedFiles/wwwpepcocom/DCComprehensiveReliabilityPlan%281%29.pdf.

[39] CDEEP (Connecticut Department of Energy & Environmental Protection). 2012. *State Vegetation Management Task Force Final Report*. August. http://www.ct.gov/deep/lib/deep/forestry/vmtf/final_report/svmtf_final_report.pdf.

[40] TPUC (Texas Public Utilities Commission). 2009. *Cost-Benefit Analysis of the Deployment of Utility Infrastructure Upgrades and Storm Hardening Programs*. March. http://www.puc.texas.gov/industry/electric/reports/infra/Utlity_Infrastructure_Upgrades_rpt.pdf.

[41] DOE (U.S. Department of Energy). 2010. *Hardening and Resiliency: U.S. Energy Industry Response to Recent Hurricane Seasons*. Washington D.C.: DOE. August. https://www.oe.netl.doe.gov/docs/HR-Report-final-081710.pdf.

[42] MCUSPR (Moreland Commission on Utility Storm Preparation and Response). 2013. Final Report. http://www.governor.ny.gov/sites/governor.ny.gov/files/archive/assets/documents/MACfinalreportjune22.pdf.

[43] MEA (Maryland Energy Administration). 2012. Weathering the Storm: Report of the Grid Resiliency Task Force. September 24, 2012. http://energy.maryland.gov/reliability/.

[44] MEA (Maryland Energy Administration). 2012. Weathering the Storm: Report of the Grid Resiliency Task Force. September 24, 2012. http://energy.maryland.gov/reliability/.

[45] MPSC (Maryland Public Services Commission). 2012. In the Matter of the Application of Potomac Electric Power Company for an Increase in its Retail Rates for the Distrubtuion of Electric Energy. Case Jacket: Case number 9331. http://webapp.psc.state.md.us/newIntranet/Casenum/CaseAction_new.cfm?CaseNumber=9311.

[46] MEA (Maryland Energy Administration). 2012. Weathering the Storm: Report of the Grid Resiliency Task Force. September 24, 2012. http://energy.maryland.gov/reliability/.

[47] MPSC (Maryland Public Services Commission). 2012. In the Matter of the Application of Potomac Electric Power Company for an Increase in its Retail Rates for the Distrubtuion of Electric Energy. Case Jacket: Case number 9331. http://webapp.psc.state.md.us/newIntranet/Casenum/CaseAction_new.cfm?CaseNumber=9311.

[48] WH (White House). 2013. Economic Benefits of Increasing Electric Grid Resilience to Weather Outages. Washington, DC: Executive Office of the President. August. http://energy.gov/sites/prod/files/2013/08/f2/Grid%20Resiliency%20Report_FINAL.pdf.

[49] LBNL. 2015. *ICE Calculator Case Study Overview: EPB Chattanooga Distribution Automation*. Lawrence Berkeley National Laboratory.

[50] DOE. 2013. *US Energy Sector Vulnerabilities to Climate Change and Extreme Weather*. DOE/PI-0013. Washington, DC: DOE. July. http://energy.gov/sites/prod/files/2013/07/f2/20130716-Energy%20Sector%20Vulnerabilities%20Report.pdf.

[51] Sussman, F., C. P. Weaver, and A. Grambsch. 2014. "Challenges in Applying the Paradigm of Welfare Economics to Climate Change." *Journal of Benefit-Cost Analysis* 5: 347–376. doi:10.1515/jbca-2014-9001. December. http://journals.cambridge.org/download.php?file=%2F4494_076C672DA551FD9E95B5407C03B4B08B_journals__BCA_BCA5_03_S2194588800000890a.pdf&cover=Y&code=a794327733e99d79559d97c60a9e146d.

[52] DOE. 2015. *Climate Change and the U.S. Energy Sector: Regional Vulnerabilities and Resilience Solutions.* Washington, DC: DOE. October. http://energy.gov/sites/prod/files/2015/10/f27/Regional_Climate_Vulnerabilities_and_Resilience_Solutions_0.pdf.

# 7. BUILD A PORTFOLIO OF RESILIENCE MEASURES

Following a preliminary assessment of identified resilience measures to address climate-related risks, power system planners and stakeholders will need to determine the most appropriate measures for inclusion in a final portfolio or action plan. This selection process will require more rigorous evaluation of the candidate measures, including comparison of refined cost/benefit estimates to specified criteria and an assessment of each measure's feasibility, efficacy, co-benefits, and ability to withstand a range of climate impacts. Resulting benefits will vary with asset and system conditions, the timing of implementation, the timing of projected impacts, the probability of climate impacts, and the collective mix of selected measures.

## 7.1 EVALUATE AND PRIORITIZE RESILIENCE MEASURES

Building an effective portfolio of resilience measures requires planners to balance multiple

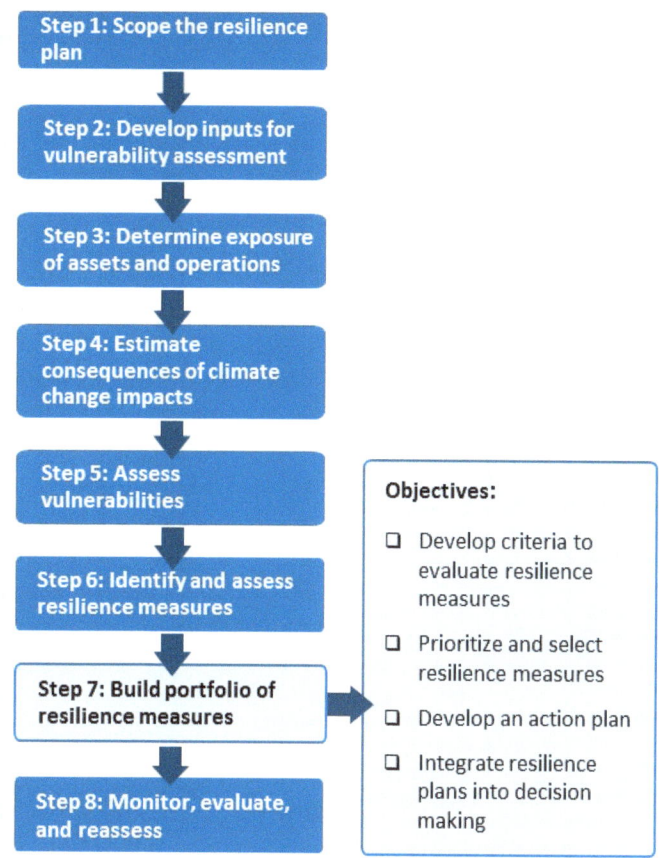

considerations and assess the tradeoffs among priority selection criteria. Beyond estimated costs and benefits, portfolio development can be heavily affected by stakeholder input, societal management objectives, resource availability (natural, human, and financial capital), and other factors. There is no single or best set of resilience measures for maintaining a resilient power supply in the face of changing climate conditions; each portfolio supports a unique utility.

## COST-BENEFIT ANALYSIS

As utilities develop their climate resilience strategies and solutions, a key step in the analytical process involves evaluating the costs and benefits of potential resilience improvements. Utilities frequently use cost-benefit analyses (CBAs) to make investment decisions. Most utilities are required to demonstrate that identified resilience projects will yield net benefits for their customers. Using the cost and benefit information discussed in Chapter 6, utilities can rank resilience measures from most to least benefit delivered per unit cost. Even if the costs and benefits cannot be quantified, a qualitative (e.g., categorized into high, medium, and low), relative comparison can help with the prioritizing of those climate resilience measures with the greatest benefit that exceeds the cost.

In any analysis, more data can always be gathered, more costs or benefits quantified, more estimates refined, and more tools used. One characteristic of a suitable CBA is the efficient use of resources: more effort should be spent on an analysis only when that effort produces a more robust result or an outcome that resonates with decision makers.

In some aspects of an analysis, nothing more than general estimates may be needed (such as the magnitude of system impacts or maintenance costs). In other cases, refining the cost-benefit analysis may be imperative—as when upfront financial costs vary across resilience measures and those costs are critical to the bottom line. Once utilities understand the criteria that define the characteristics of different resilience measures, these criteria can be considered in combination to construct a preferred portfolio of resilience measures that will meet the goals of the asset owners and stakeholders.

Tools for visualizing comparisons and interactions among measures may enhance understanding of the relative costs and benefits, facilitating selection of a portfolio of measures. For example, utilities can evaluate the ability of a specified portfolio of risk reduction measures to deliver improved system performance against a variety of metrics. Such analyses can help utilities understand when resilience investments begin to yield diminishing returns. As shown in Figure 13, plotting the cost of customer outages against the costs of investments that improve system resilience by improving CR-90 (i.e., reducing the time needed to restore power to 90% of customers after a severe storm [see Chapter 6]), can help determine the point at which resilience investments begin to show diminishing returns. Stakeholder and expert input may be needed to augment or refine visualizations to suit the local conditions and decision context.

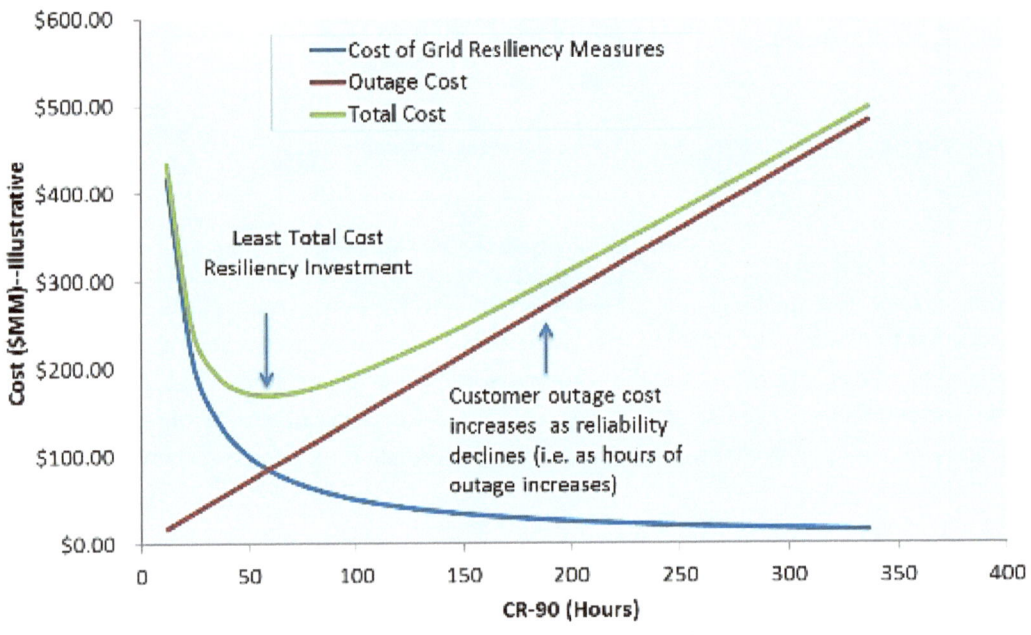

*Note: Outage Cost likely not linear and will vary by customer class.*

**Figure 13.** Example visualization of a total cost analysis of grid resilience measures.[1]

**Case Study: Using Cost-Benefit Ratio to Compare Potential Resilience Measures**

Entergy Corporation has developed a framework and undertaken a study to quantify climate risks and help inform approaches for building a resilient U.S. Gulf Coast. Entergy found that the Gulf Coast is vulnerable to growing environmental risks today and faces an estimated $350 billion or more in cumulative losses by 2030. Key uncertainties involved in addressing this vulnerability include the impacts of climate change, and the cost and effectiveness of resilience measures.

The study covers coastal counties and parishes on a strip of land stretching up to 70 miles inland across the shoreline of southern Texas, coastal Mississippi, and Alabama. This area is threatened by hurricanes, which typically cause damage primarily through extreme winds, storm surge, and flooding. In order to calculate costs and benefits, the study considered the costs of all damaged assets and interrupted business activity in the study region.

Recognizing the uncertainty of potential loss aversion, the study identified potentially attractive measures that yield a cost-to-benefit ratio of less than 2. The measures are compared on an overall cost curve, in which the width of each bar represents the total potential of that measure to reduce expected losses to 2030 for a given scenario, and the height of each bar represents the ratio between costs and benefits for that measure. Along with $76 billion in private funding, approximately $44 billion of public funding will be required over the next 20 years to fund key infrastructure projects. [2]

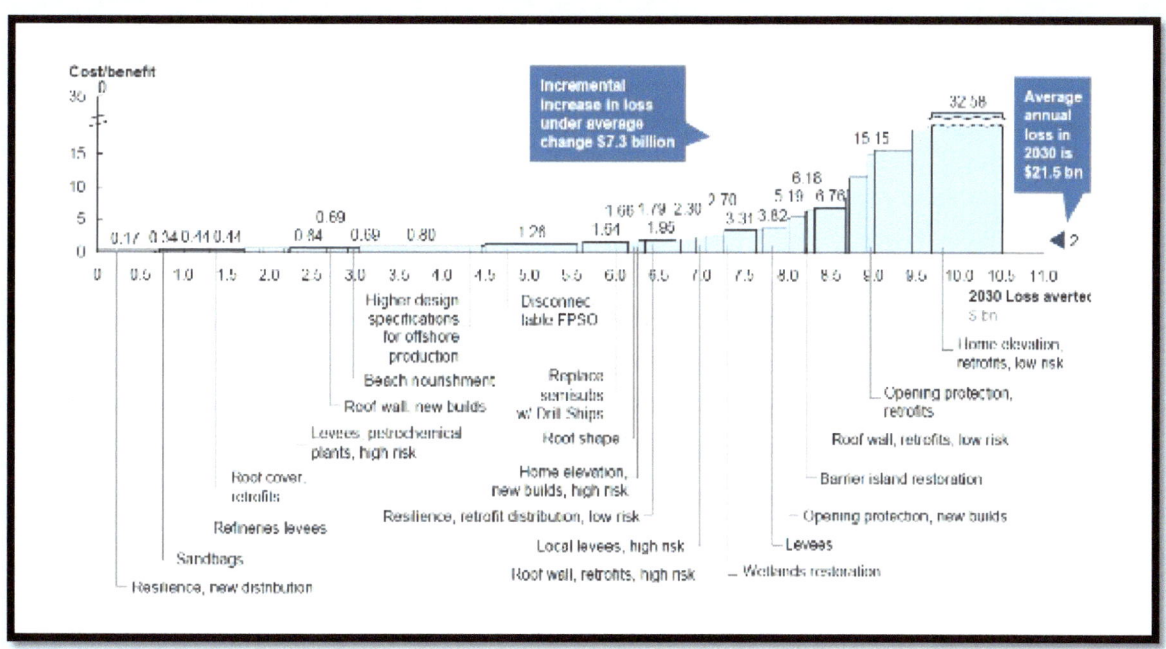

**Figure 14.** Marginal cost (in ratio of benefits to costs) of each identified resilience measure. [3]

Cost-benefit analyses must be applied thoughtfully, because frameworks that apply to reliability projects are not always adequate for planning resilience projects. For example, up-front costs for new construction may appear high in a CBA if the benefits achieved are spread over a long lifetime. Investments in resilience may coincide with multiple other planning goals (e.g., capacity expansion or replacement). A cost-benefit analysis that only compares the high costs of a multi-purpose project to the benefits of increased resilience will return incomplete results.[4] The discount rate applied to future costs and benefits is a critical assumption with no clear solution—the assumption can tip the balance between a measure being considered favorable or unfavorable in a CBA.[5] When key variables, such as projected climate events, costs, or outage duration are unknown or cannot be reliably estimated, utilities should consider a variety of evaluation approaches, including sensitivity analysis, breakeven analysis, balanced scorecard analysis, or robust decision making (see box).

**Approaches that may be useful for dealing with uncertain information and resilience investments**

**Sensitivity analysis** can be used to determine the importance of certain variables (i.e., climate events, cost, or other uncertain data) on the outcome and results.

**Breakeven analysis** can be used to help determine the value of each investment in cases where the probability of severe weather and the probability of damage to infrastructure are difficult to quantify. In one application of the method, the benefit of proposed investments is calculated by estimating the number of customer outages during climate or extreme weather events that the investments would mitigate and estimating the value that customers place on avoiding extended outages during such events. The "breakeven" point is where the value of lost load is equivalent to the cost of the investment.[6,7]

**Balanced scorecard analysis** incorporates non-cost information with cost metrics to provide a more 'balanced' evaluation. The approach is intended for use as a management system (not only metrics) to clarify company strategy and translate measurement into action.

**Robust Decision Making (RDM)** may be implemented using relatively simple approaches that array the options and results under a range of climate futures and display the information in ways that are relevant for decision makers. In more sophisticated RDM applications, weights can be assigned for stakeholders and decision-makers, and more complex mathematical algorithms are used to obtain the results.[8] RDM is particularly useful in situations with high uncertainty.

When selecting effective and appropriate resilience strategies, another key consideration is the lifetime of the infrastructure versus the severity of projected climate impacts. For example, some electricity assets may have a relatively short design or use lifetime, which may suggest that it would be appropriate to monitor conditions over the near-term and be prepared to recommend changes in asset composition as needed. A power generation facility, on the other hand, has a long service life, so the design should logically anticipate future threats. Some resilience measures may support a short-term objective, while others may persist over an extended period. The planning horizon for resilience measures should be consistent with the lifetime of the infrastructure.

Use of multiple criteria to evaluate resilience measures will help to inform construction of a robust portfolio. While many utilities' decision processes are built around a traditional CBA, additional metrics (as described in Chapter 6) may significantly complement the CBA information. Metrics, qualitative or quantitative, that relate to the robustness, timeliness, and flexibility of the resilience measure should be considered (see example criteria in box below).

**Evaluating Resilience Measures: Additional Criteria to Augment a CBA**

**Co-Benefits:** Positive impacts of climate resilience measures beyond energy sector resilience. A risk management measure might provide additional benefits, in addition to reducing the specific climate-related risks of concern. Co-benefits may include positive economic impacts on other sectors, reductions in GHG emissions, improved health and security of vulnerable populations, or benefits to ecosystems.

**Robustness:** The anticipated performance of a risk management measure under a wide range of possible climate futures. It may be relatively costly to select an option that is more robust, so the incremental cost vs benefits of additional robustness may need to be considered.

**Effectiveness:** A measure of how well the risk management measure reduces the specific climate risks of concern and generates the primary benefits sought (e.g., damages reduced, costs avoided, lives saved) over an appropriate time horizon. The decision maker may specify benefits categories to help define effectiveness.

**Reversibility and Flexibility:** The extent to which a measure can adapt to, be revised, or be reversed in response to changing conditions, needs, or regulatory requirements. Flexibility may be an especially important consideration for measures that are long-lasting, are relatively costly, and/or have irreversible consequences.

**Rapidity:** The speed with which disruptions can be addressed and safety, services, and financial stability restored is critical, particularly for operations manager dealing with climate impacts and extreme events. The measure could be applied to structural solutions, operational actions to mitigate damages, or the dissemination of advanced warning, guidance, and resources to vulnerable populations.[9]

**Case Study: Consolidated Edison (ConEd) Prioritization of Resilience Measures**

To address issues related to climate change and severe weather, ConEd conducted a process for prioritizing storm hardening solutions. The process was designed to realize the greatest benefits compared to costs and facilitate rapid implementation. The prioritization process considered factors such as public safety, population impact, critical infrastructure reliance on the electric system, the vulnerability of the systems, and the investments needed to achieve hardening.

ConEd identified several strategies based on recent experience or recommended by commissions that Governor Cuomo established after Superstorm Sandy. The strategies included undergrounding and flood protection projects, including floodwalls for certain electric and steam equipment, raising critical equipment above potential flood levels, and accelerating installation of submersible equipment, as appropriate.

The company evaluated 14 substations and six power generation facilities that were impacted by Sandy and plans to evaluate other facilities not directly impacted. The evaluation found the following equipment most susceptible to flooding: relay houses, control panels, control rooms, diesel generators, AC and DC power supplies, and pumping plants. Protective measures include: elevating equipment; enhancing seals around connections; preemptively de-energizing non-operationally critical equipment to protect against control/power supply short circuits; installing flood barriers, watertight doors, sluice gates, and flood pumps to prevent the migration of water into the stations; eliminating facilities by converting the local distribution system to 13kv or 27kV autoloops; and using fiber optic-based communications and control to provide more effective fault protection during flooding .

To optimize overall risk reduction, a mix of solutions was proposed at varying levels of program spending across substations and transmission and distribution networks. ConEd also proposed to improve the flexibility of the electric distribution system, including the installation of additional switches and related smart grid technology and the reconfiguration of certain networks to reduce the impact to customers most affected by certain storms. [10]

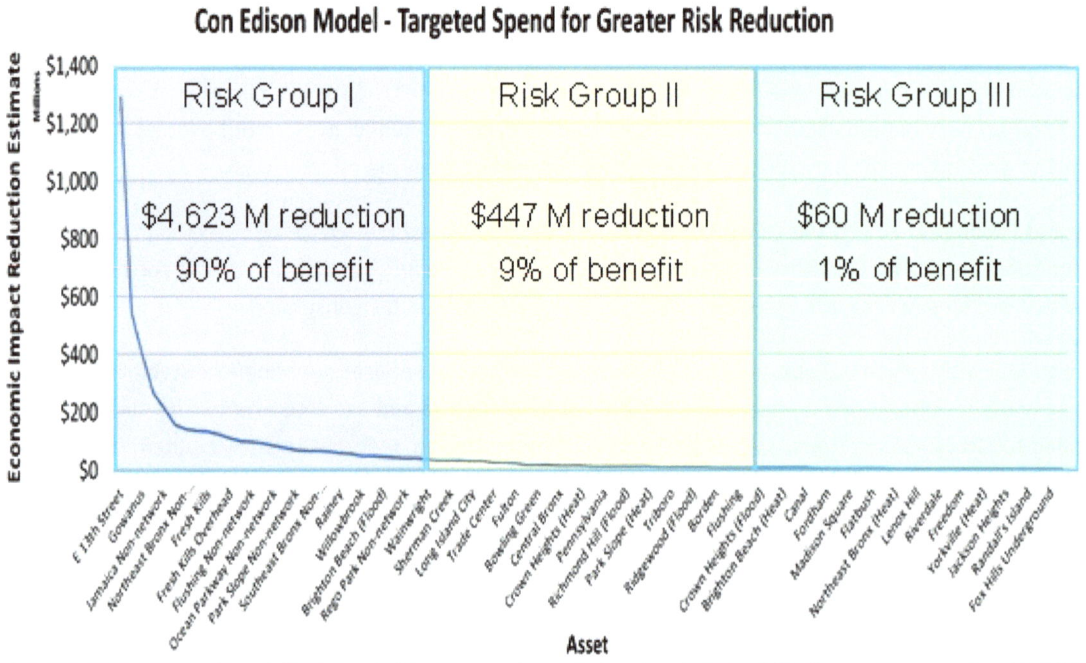

Figure 15. ConEd risk prioritization results. [11]

**Breakeven Analysis of PSE&G's Energy Strong Program**

In February 2013, Public Service Electric & Gas Company (PSE&G) submitted their proposal for the company's Energy Strong program to the New Jersey Board of Public Utilities. In support of this proposal, a breakeven analysis was applied to the Energy Strong program by the Brattle Group. The breakeven approach was introduced as an alternative method for evaluating resilience investments; essentially avoiding the need to estimate the probabilities of severe weather events and uncertainties associated with the impacts of such events.

In the analysis, the value of the investment is given in minutes of customer interruption (CMI) that could be mitigated over the lifetime of the investment. The "breakeven" point is the interruption time that has value of lost load equivalent to the cost of the investment. The breakeven point can be defined as $E(B) - C = 0$, where $E(B)$ are expected benefits and C is the predetermined cost of the investment. $E(B)$ is the probability of a climate or weather event multiplied by the benefit of resilience investment in term of outage minutes avoided. If C is known, then $E(B)$ or VOLL multiplied by unserved kWh can be used to determine the minutes of outage to "pay-back" the resilience investment. The value of the investment can then be compared to historical outage data and the probabilities of outages associated with climate risks in the future to the breakeven number of outage minutes to assess the expected benefits of the investment.

Through this analysis, it was determined that the proposed Electric Energy Strong program would result in reductions in the number and duration of outages caused by severe weather events, providing value to customers. The analysis found that this value is equal to the cost of the proposed Electric Energy Strong program for cumulative outage durations of three days. Either through a single major future weather event, such as another Hurricane Sandy, or from the combination of lesser weather events taking place over the course of the life of the Electric Energy Strong assets, customers would realize the value of the investments. The Brattle Group did not evaluate in this study the co-benefits to society from resilience improvements. [12]

## 7.2 DEVELOP A RESILIENCE ACTION PLAN

Ultimately, selecting the right mix of resilience measures can be challenging, even after conducting an objective prioritization process (as discussed in the previous section). A resilience action plan specifies which risks to address, how to address them, and when. To facilitate the prioritization of resilience investments, action plans should clearly articulate the utility's core objectives and define its overall vision of resilience. By listing the resilience measures to be implemented, the plan will implicitly define what is deemed an unacceptable level of climate risk.

For some resilience measures, the challenge is not to determine whether the measure is needed, but at what point the utility should act. Utilities should consider implementing selected options in distinct phases. This approach lets utilities learn lessons during initial phases that may save time, money, or resources later. Gathering feedback after each phase and incorporating it into an evolving plan may also improve efficiency and effectiveness. In addition, a phased approach improves flexibility, in case priorities change over time. Similarly, utilities might consider running pilot programs before attempting larger implementation projects. [13]

Long-term planning horizons should be incorporated into the action plan; resilience planning cannot be done well in the five-year increments often used for infrastructure planning. As part of the long-term planning process, utilities should look for opportunities to incorporate resilience measures into scheduled replacements or upgrades, thus accelerating resilience improvements in a cost-effective manner (see box: AVANGRID Resilience Actions Incorporated with System Planning). In addition, planning processes can facilitate the installation of more resilient infrastructure during repair and restoration activities after severe events. Understanding and planning for the implementation of priority resilience measures will allow asset owners to rebuild strategically and far more cost-effectively than in reaction to damaging events.

Electric utilities will be well served to track the actual costs and measure the effectiveness (if tested by a climate or weather event) of each action and make any adjustments necessary to the evaluation of resilience options. If actions are not producing anticipated outcomes, planners may consider modifying the evaluation approach or correcting the action plan. As necessary, deliberations in prior steps can be revisited. With hindsight, planners may be able to spot an oversight or miscalculation. If so, they should review the options, re-evaluate risks, and then decide whether additional and/or new actions are needed. Utilities should continue to iterate in a process of continual improvement as new information becomes available.

**Case Study: AVANGRID Resilience Actions Incorporated with System Planning**

AVANGRID identified resilience measures and strategies by reviewing vulnerable assets and operations for three of the companies (RG&E, NYSEG, and Central Maine Power). In one example, the company characterized 'increase in temperature and heat waves' as a threat that would increase customer demand while the electric line and substation equipment rating would decrease. In its vulnerability assessment, the company noted that higher ambient temperatures, especially over a prolonged period, could have a significant impact on system load. If the average temperature increases by 5°F by 2050, peak ambient temperatures could increase such that facility ratings would have to be decreased to maintain proper conductor sag clearances to comply with National Electric Safety Code requirements. AVANGRID plans to manage these risks by incorporating upgrades, as needed, to correspond with existing system planning cycles and methods. This provides an effective and cost-conscious method for making resilience improvements to address threats that are gradual. Relevant resilience strategies identified include:

- Adjusting facility load limits for higher ambient temperature conditions.
- Installing reclosers, and possibly sectionalizers and circuit breakers along with automatic fault detection and sectionalizing intelligence that allow better monitoring and control of the system and improve restoration time.
- Building in redundancy to tie distribution lines together to allow back feeding of circuits.
- Incorporating microgrids and self-healing sub-transmission and distribution systems and non-transmission alternatives to reduce transmission line loading.
- Pursuing demand response, energy efficiency and localized alternatives to traditional infrastructure construction[14]

## INTEGRATE RESILIENCE ACTION PLANS INTO CORPORATE DECISION MAKING

Incorporating resilience action plans into existing processes or scheduled plan updates can be an effective way to expedite action. Using the action plans in this way may be seen as improving existing analysis and practice, rather than as a separate and distinct activity. Considering climate change as one of many risks to be evaluated in corporate decision-making rather than as a separate issue can significantly lower barriers to implementation.

Utilities may be able to incorporate resilience action plans into existing processes, such as the asset management process. Asset management is a natural fit as a way to incorporate most resilience actions and information. Replacement or restoration of assets to improve resilience can also be integrated into emergency management, hazard mitigation plans, planning project selection criteria, or environmental reviews.

# CHAPTER 7 REFERENCES

[1] DOE (U.S. Department of Energy). 2016. *Climate Change and the Electricity Sector: Guide for Assessing Vulnerabilities and Developing Resilience Solutions to Sea Level Rise*. Draft, April 2016.

[2] Entergy Corporation. 2010. *Building a Resilient Energy Gulf Coast: Executive Report*.
www.entergy.com/content/our_community/environment/GulfCoastAdaptation/Building_a_Resilient_Gulf_Coast.pdf.

[3] Entergy Corporation. 2010. *Building a Resilient Energy Gulf Coast: Executive Report*.
www.entergy.com/content/our_community/environment/GulfCoastAdaptation/Building_a_Resilient_Gulf_Coast.pdf.

[4] DOE (U.S. Department of Energy). 2016. *Climate Change and the Electricity Sector: Guide for Assessing Vulnerabilities and Developing Resilience Solutions to Sea Level Rise*. Draft, April 2016.

[5] EPA. 2010. *Chapter 6: Discounting Future Benefits and Costs*. Chapter in *Guidelines for Preparing Economic Analyses*. Washington, DC: U.S. Environmental Protection Agency. https://yosemite.epa.gov/ee/epa/eerm.nsf/vwan/ee-0568-06.pdf/$file/ee-0568-06.pdf.

[6] Fox-Penner, P., and W.P. Zarakas. 2013. Analysis of Benefits: PSE&G's Energy Strong Program. October.
http://www.brattle.com/system/testimonies/pdfs/000/000/936/original/Analysis_of_Benefits_-_PSE_G's_Energy_Strong_Program_Fox-Penner_Zarakas_10_07_13.pdf?1399065260.

[7] Zarakas et al. Utility Investments in Resilience: Balancing Benefits with Cost in an Uncertain Environment. 2014
http://dx.doi.org/10.1016/j.tej.2014.05.005.

[8] Lempert, R. 2015. Embedding (some) Benefit-Cost Concepts into Decision Support Processes with Deep Uncertainty. *Journal of Benefit-Cost Analysis*. Vol. 5 (3), pp. 487–514. http://www.degruyter.com/view/j/jbca.2014.5.issue-3/jbca-2014-9006/jbca-2014-9006.xml?format=INT.

[9] DOE (U.S. Department of Energy). 2016. *Climate Change and the Electricity Sector: Guide for Assessing Vulnerabilities and Developing Resilience Solutions to Sea Level Rise*. Draft, April 2016.

[10] ConEd. 2013. Consolidated Edison of New York. Storm Hardening and Resiliency Report. December 4, 2013.
https://s3.amazonaws.com/nyclimatescience.org/Storm%20Hardening%20and%20Resiliency.pdf

[11] ConEd. 2013. Consolidated Edison of New York. Storm Hardening and Resiliency Report. December 4, 2014.
https://s3.amazonaws.com/nyclimatescience.org/Storm%20Hardening%20and%20Resiliency.pdf

[12] Fox-Penner, P., and W.P. Zarakas. 2013. Analysis of Benefits: PSE&G's Energy Strong Program. October.
http://www.brattle.com/system/testimonies/pdfs/000/000/936/original/Analysis_of_Benefits_-_PSE_G's_Energy_Strong_Program_Fox-Penner_Zarakas_10_07_13.pdf?1399065260.

[13] USCRT (U.S. Climate Resilience Toolkit). 2016. Evaluating Risks & Costs. https://toolkit.climate.gov/get-started/step-4-prioritize-actions#4.3.

[14] AVANGRID 2016. *Assessment of Vulnerabilities Due to Climate Change and Extreme Weather*. Iberdrola USA Networks. January.

Planning processes by their nature contend with uncertainty about the future. Planning for climate change and extreme weather hazards includes uncertainty not only about how the climate will differ in 10, 30, or 50 years, but also about how technologies, consumer demand, and policies affecting the energy sector may change in parallel. Planning processes should also account for uncertainty about how different business units incorporate and act upon climate resilience action plans. A robust plan should be adaptable to changing expectations and evidence, as well as facilitate monitoring of progress and evaluation of implemented actions. A robust plan necessarily incorporates the following:

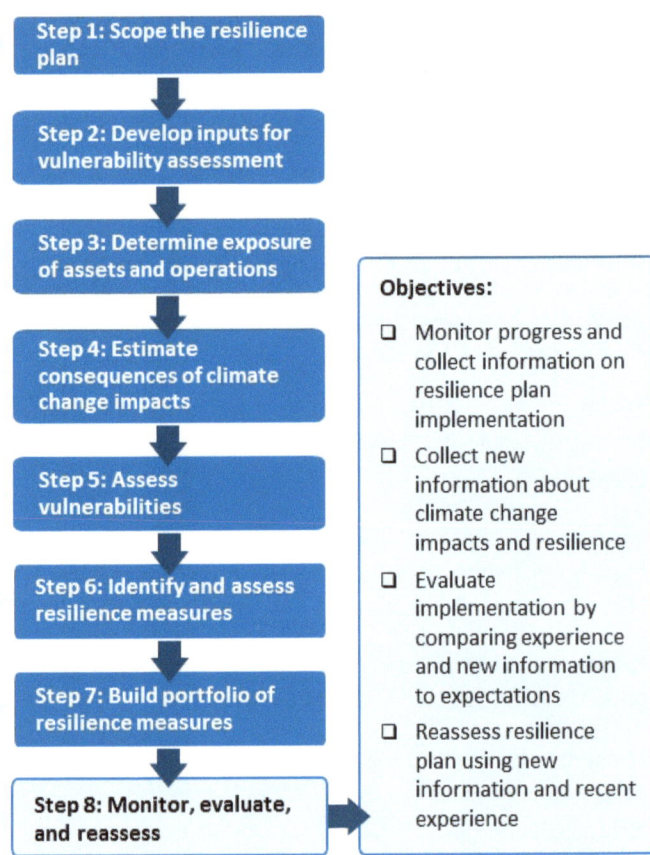

**Objectives:**

☐ Monitor progress and collect information on resilience plan implementation

☐ Collect new information about climate change impacts and resilience

☐ Evaluate implementation by comparing experience and new information to expectations

☐ Reassess resilience plan using new information and recent experience

- **Monitoring progress**—measuring implementation milestones against the resilience plan and gathering feedback from other business units implementing the resilience plan
- **Evaluation of implementation**—assessing the effectiveness of completed resilience actions and incorporating and comparing feedback with new information about climate change, consumer demand, energy policies, installation costs, resilience technologies, and implementation experience
- **Reassessing the plan**—reassess the vulnerability assessment and resilience plan by completing a periodic review or by repeating steps when new information becomes available

## 8.1 MONITOR PROGRESS

Once a utility has completed and obtained approval for its resilience plan, executing the plan will likely require coordination across multiple business units and skill sets. Due to differences in organization, facilities, business activities, and operating procedures, utilities may handle implementation in different ways. However, monitoring implementation progress should be a central aspect of any implementation process. As each important stage of the resilience plan is completed ("implementation milestones"), planners should take the opportunity to collect feedback about the process.

Implementation milestones are key points in the resilience plan implementation process that indicate an increased level of resilience to a specific climate threat has been achieved. Milestones can include the completion of construction for asset hardening or relocation measures. For operational resilience measures, milestones may include a percentage of staff or facilities who have received updated training or which have initiated updated procedures.

The selection of milestones should be suited to the needs of a utility. For example, utilities with multiple sites facing many vulnerabilities may wish to introduce resilience measures as pilot programs that can be reevaluated before wide-spread adoption. Utilities with particularly high-risk vulnerabilities may wish to implement the highest-priority resilience measures first.

As a utility achieves implementation milestones, it is important to monitor and collect key cost and performance data that can be used to evaluate the implemented actions. Critical cost data will include not only the total costs of upgrades, installations, and other direct expenditures, but also financing costs, and planning and construction lead times. Important performance data should include performance on metrics important for increasing system resilience (e.g., CAIDI/SAIFI, safe operating temperature, and degrees of redundancy) and how resilience upgrades affect other system performance (i.e., beneficial or adverse effects unrelated to climate and extreme weather resilience), as well as how well metrics for assessing these performance data perform.

## 8.2 EVALUATE IMPLEMENTATION

Once new information is collected from monitoring implementation, this data should be evaluated against expectations and assumptions used in the vulnerabilities assessment and resilience plan. Where possible, data collected from real-world experience should be evaluated side-by-side with model inputs used for assessing the costs and benefits of resilience measures. New costs or benefits which have not previously been estimated should also be included in the evaluation. Evaluation questions to ask at this point may include:

- Do resilience actions meet or exceed expected costs?
- Do the resilience improvements achieve expected reductions in vulnerability to climate threats?
- Are improvements in system performance and reliability achieved?
- Are there any new, unanticipated costs or benefits that arise as a result of a resilience action?
- Are the metrics being used the best available for identifying cost and benefits (e.g., VOLL definition)?

Evaluating implementation should also take into account new information from outside sources. One of the most important types of new outside information is updated climate change science or projections, especially updates to the major assessment literature. New information can also include new tools for understanding and evaluating vulnerabilities, new reports or case studies on resilience technologies or options, new data on resilience measure costs, and any other relevant information that may affect the results of the vulnerabilities assessment and resilience plan. Major sources for updates to climate science and projections include:

- Intergovernmental Panel on Climate Change assessment reports and products
- U.S. Global Change Research Program Reports
- DOE Reports on climate change resilience planning
- NOAA climate change projections

## 8.3 REASSESS THE PLAN

Resilience plans should be reassessed in order to incorporate both feedback from implemented resilience actions, as well as updated information about climate change, resilience technologies and planning tools, or connected infrastructure vulnerabilities. Reassessing the vulnerabilities assessment and resilience plan should be a regular part of the planning process that can occur in several different ways, depending on how new information becomes available, the urgency or degree of difference presented in new information, or the resource constraints of the utility's resilience planning process.

Regular periodic updates to the resilience plan are a good approach to incorporate new climate change information that is constantly being produced. Periodic updates also present a good opportunity to systematically review the experience of all business units implementing a resilience plan. Regular updates should occur at least as frequently as major climate change assessment reports are produced by the IPCC or USGCRP, since these assessment products will include the most certain updates to rapidly developing areas of climate science.

When an evaluation of feedback from implementation experience or other new information demonstrates that the outcome or conclusions of a resilience plan may be affected, individual steps of the resilience plan can be repeated, and decisions based on the outcome can be individually updated. When reassessing an individual element of the resilience plan, it is important to begin with the step most likely to be affected by the new information, although this may not always be immediately clear. Each step of the vulnerabilities assessment and resilience planning process may be affected by either new information or by feedback on implementation.

Some hypothetical examples of how utilities could monitor progress, evaluate implementation, and reassess their resilience plans follow:

- **Example:** As part of its ongoing resilience planning process, a utility commences a wholesale update to its resilience plan completed more than one year prior.
    - o **Monitor Progress:** The utility has been collecting cost and performance data on all of its resilience-building investments and operational changes.
    - o **Evaluate Implementation:** Incorporating the findings of the newest IPCC and USGCRP assessment reports, the utility finds that projections for elevated temperatures in their region may raise concerns for substation and transmission line capacity, a type of climate threat that was not considered in the scope of the previous resilience plan.
    - o **Reassess Resilience Plan:** Using the updated climate projections, as well as the lessons learned from earlier implementation efforts, the utility starts with Chapter 1 of the resilience planning process, and expands the scope to meet include new climate risks.
- **Example:** A utility with multiple substations within the inundation zone for a Category 2 hurricane identified in its resilience plan the installation of floodwalls as the most cost-effective means of flood protection using cost-benefit analysis.
    - o **Monitor Progress:** After beginning constructing on a floodwall at the most critical substation, the utility experiences geotechnical problems with the soils underlying the floodwall, leading to higher construction costs than were initially anticipated, as well as significant construction delays.
    - o **Evaluate Implementation:** The utility compares the real construction costs to the assumptions in its resilience plan and discovers that depending on the conditions at other sites, alternative methods of protecting substations may be more cost effective.

- **Reassess Resilience Plan**: The utility decides to revisit Steps 6 and 7 of the plan, this time incorporating updated details from the remaining substation sites, and determines that in some sites, floodwalls would likely not be the lowest-cost resilience option. The utility decides instead to install submersible equipment or elevate existing equipment at these sites.

- **Example:** In its resilience plan, a utility identifies transmission lines vulnerable to wildfire and decides on an accelerated vegetation management schedule to reduce its exposure to wildfire hazards.
    - **Monitor Progress:** After implementing a schedule that doubles the frequency of brush and tree-clearing visits, the utility finds that after a year, crews are not clearing nearly as much vegetation on each pass through a right-of-way.
    - **Evaluate Implementation:** The utility compares the reports from work crews with the assumptions in their resilience plan about vegetation management effectiveness and finds a sizeable disparity.
    - **Reassess Resilience Plan:** With updated information about the effectiveness of vegetation clearing, the utility chooses to repeat the analysis of resilience measure costs and benefits in Chapter 6 of their resilience plan. On reassessment, the utility decides a slightly less-frequent vegetation management schedule is an acceptable balance of costs and benefits.

This Guide sets forth a flexible approach to climate resilience planning that can be tailored to the unique needs, goals, and resources of specific electric utilities and electricity system operators in preparing for a range of climate change impacts and extreme weather. The Guide highlights a number of available tools, projections, sample metrics, and completed assessments that are now available to assist and guide planners in identifying risks, evaluating options, and developing effective plans. While significant gaps in tools, resources, and methodologies remain, this flexible framework paves the way for planners and decision makers across the country to immediately move forward in developing and implementing plans to make their electricity systems more resilient to projected climate impacts. Early planning and action, such as integrating climate resilience considerations into regular planning processes and system maintenance decisions, can ultimately reduce the significant costs of climate change to U.S. electricity systems, their service areas, and the national economy.

Ongoing efforts to fill the existing gaps in data, methodologies, tools, and other resources are underway at the U.S. Department of Energy and its National Laboratories, as well as numerous academic, government, and industry organizations across the country. Continued communication, sharing, and coordination of needs, research, and solutions will help leverage resources and accelerate progress and resilience on all fronts. Current research and development to improve vulnerability assessment and resilience planning practices focus on the following objectives:

- Improve the collection, organization, and availability of actionable data relevant to climate resiliency planning with an appropriate temporal and spatial resolution.

- Develop and standardize advanced metrics that are specifically designed to capture unique facets of climate resilience.

- Develop, update, expand, and refine tools to determine the costs and benefits of climate resilient solutions.

- Develop and deploy clean, affordable, and reliable energy technologies that significantly enhance climate resilience and preparedness

- Establish enabling policy frameworks to incentivize and accelerate investment in climate resilience.

The *Quadrennial Energy Review* identifies an urgent need for better data, metrics, and analytical frameworks to help build resilience, reliability, and security in the energy sector.[1] DOE notes that gaps in actionable data are impeding investment in and decision-making on resilience. Expanding the availability of data is essential to assist decision makers in effectively evaluating risks to infrastructure and making informed investments in resilience.

The *Quadrennial Energy Review* suggests that "DOE, in collaboration with DHS and interested infrastructure stakeholders, should develop common analytical frameworks, tools, and metrics for assessing the resilience, reliability, and security of energy infrastructures."[2] Access to such resources will help electricity system planners and decision makers identify, prioritize, and justify appropriate investments in system resilience and will complement ongoing federal activities to provide data, information, and tools through the Climate Data Initiative and the Climate Resilience Toolkit (see Section 2.1: Develop Inputs on Climate Change). The lack of broadly accepted resilience planning frameworks impedes efforts by electricity companies to compare or leverage the planning outputs of other companies with similar systems or challenges. As the DOE's *Quadrennial Energy Review* suggests, common analytical frameworks, tools, and metrics would assist utilities in planning, setting priorities, and justifying expenditures for climate resilience.

In addition, current tools and metrics face limitations in their ability to accurately identify the costs and benefits of specified resilience projects. For example, most measures of VOLL, such as and ICE Calculator, are not designed to estimate costs associated with long-term outages, limiting their suitability for evaluating resilience projects. In the absence of good metrics to evaluate resilience investments, some utilities are instead using reliability metrics. Unfortunately, reliability metrics typically fail to reflect the full benefits of these investments. Reliability metrics and tools primarily measure "blue-sky" conditions, not extreme weather conditions. In addition, they primarily rely on customer surveys following short-term outages, not the long-term outages that tend to follow extreme weather events. DOE, other federal agencies, the private sector, and other organizations are continuing to address these critical challenges in quantifying the true costs of climate impacts and the full benefits of resilience improvements.

Establishing a more climate-resilient energy sector will require improved technologies and supportive policies for timely deployment. The current electricity infrastructure was designed to operate under past environmental conditions, which are shifting with changing climate patterns. Research, development, demonstration, and deployment (RDD&D) efforts can generate cost-effective energy technologies to replace our aging electricity infrastructure while simultaneously building resilience to risks posed by climate change and extreme weather. A policy framework that fosters climate resilience would broaden the suite of advanced technologies available in the future, and strong deployment policies could address existing market failures by improving the cost effectiveness of climate resilience actions. Of course, decision makers will never have complete information, so "no regret" or flexible strategies that allow mid-course corrections may lead to greater and more efficient resilience in the short and long term. Ultimately, the development and deployment of climate-resilient energy technologies will build a more resilient U.S. electricity system, create new domestic and global markets, and provide greater climate resilience both nationally and worldwide.

Building a U.S. electricity sector that is more resilient to climate change and extreme weather will require coordinated and collaborative efforts across government, academia, and the private sector. While the U.S. energy sector is primarily owned and operated by the private sector, DOE can contribute by focusing its extensive RDD&D and policy expertise and capabilities on finding solutions to these complex and pressing challenges.

## CHAPTER 9 REFERENCES

[1] DOE. 2015. *Quadrennial Energy Review: Energy Transmission, Storage, and Distribution Infrastructure*. Washington, DC: U.S. Department of Energy. April. http://energy.gov/sites/prod/files/2015/04/f22/QER-ALL%20FINAL_0.pdf.

[2] DOE. 2015. *Quadrennial Energy Review: Energy Transmission, Storage, and Distribution Infrastructure*. Washington, DC: U.S. Department of Energy. April. http://energy.gov/sites/prod/files/2015/04/f22/QER-ALL%20FINAL_0.pdf.

## APPENDIX A: CLIMATE CHANGE SCENARIOS

Climate change scenarios are useful in characterizing future climate changes and in comparing the hazards projected under different assumptions. Fundamentally, climate change scenarios encode a set of assumptions about future greenhouse gas (GHG) emissions, atmospheric concentrations, or warming rates, and allow different modeling teams to compare their results on a common basis. Integrated Assessment Model (IAM) scenarios may also include assumptions about economic, population, and technology changes affecting future emissions. The scenarios used most commonly in climate models today reflect the four representative concentration pathways defined by the Intergovernmental Panel on Climate Change (IPCC) assessment reports.

**Representative Concentration Pathways (RCP):** The RCP series of scenarios are currently the most up-to-date standards used by climate modelers and were developed for and adopted by the *IPCC's Fifth Assessment Report* (AR5) in 2014.[k] Each of the four RCP scenarios is named according to the expected increase in solar radiative forcing[l] in the year 2100. Each scenario includes explicit assumptions about the amount of GHGs in the atmosphere and the global emissions trajectory to reach the final concentration in 2100. The RCP scenarios also include implicit assumptions about the population changes, economic growth, and technological deployment supporting the representative emissions trajectory. RCP scenarios, which are used in the fifth phase of the Climate Model Intercomparison Project (CMIP5), include the following:

- **RCP2.6:** A very-low emissions scenario, global GHG emissions peak early and decline over the course of the 21$^{st}$ century. Consequently, GHG concentrations and human-caused radiative forcing peak mid-century and decline by 2100 (to a level of 2.6 W/m$^2$).

- **RCP4.5:** In this low-emissions scenario, global GHG emissions stabilize by mid-century and decline rapidly thereafter. In RCP4.5, total GHG concentrations (and radiative forcing) slow by 2100 but do not stabilize until after 2100.

- **RCP6:** In this medium-emissions scenario, GHG emissions stabilize later in the 21$^{st}$ century than in RCP4.5 but also decline rapidly thereafter. In RCP6, as in RCP4.5, GHG concentrations (and radiative forcing) do not stabilize until after 2100.

- **RCP8.5:** A high-emissions scenario, GHG emissions and concentrations rise throughout the 21$^{st}$ century and do not stabilize.

**Special Report on Emissions Scenarios (SRES):** Until recently, SRES scenarios were the ones most commonly used in climate modeling. They were used in the CMIP3 ensemble, which provides the main climate projections for *IPCC's Fourth Assessment Report* (AR4). Developed in 2000, four families of SRES scenarios describe distinct socioeconomic storylines, including the following:

- **A1:** The A1 family of scenarios describes a globalized world undergoing rapid economic and population growth accompanied by the development and spread of new technologies. Sub-scenarios include versions focused on fossil fuels or renewable energy sources. Emissions in A1 scenarios are based on assumptions about which technologies are used.

---

[k] Additional information about the RCP scenarios can be found here: http://sedac.ipcc-data.org/ddc/ar5_scenario_process/RCPs.html

[l] Radiative forcing is the change in incoming solar radiative energy measured at the surface of the earth (in watts per square meter) due to increases in GHGs.

- **A2:** A2 scenarios describe a world with independent or regionally focused growth, growing populations, and overall high emissions.

- **B1:** The B1 family of scenarios shares the globalized world of A1 but with an increasing share of economic growth in services and information, reduced energy importance, lower population growth, and more global cooperation on sustainability. Emissions in the B1 family are the lowest of the SRES scenarios.

- **B2:** Like the A2 scenarios, B2 features a regionally and nationally independent world characterized by lower population and economic growth. Technological adoption is also slower.

**Other Scenarios:** Many climate projections have been generated beyond those used for the IPCC assessments. Of these, some use variations of the RCP or SRES scenarios, while others use wholly independent scenarios. Common types of scenarios include the following:

- **Emissions-based scenarios:** Many individual studies evaluate climate changes based on specific GHG limits, often in response to specific policies or goals. For example, scenarios corresponding to an 80% reduction in GHG emissions by 2050 compared to 2005 levels are sometimes used to estimate the climate outcomes resulting from national or intergovernmental commitments. When using projections based on these types of scenarios, multiple climate models should be used to avoid generating unreliable estimates as a result of regional or physical biases in any individual model.

- **Technology- or policy-oriented scenarios:** In Integrated Assessment Models, scenarios are often defined by assumptions about technologies, energy resources, or economic growth. In these types of models, climate simulations will be driven by GHG emissions generated by functions within the model. However, these types of scenarios cannot be easily compared to those of other models, which may limit their value for vulnerability assessments.

Table B.1. Potential Resilience Measures to Mitigate Climate Change and Extreme Weather Risks. [1]

| Type of Measure | Climate Threats | | | | | |
| --- | --- | --- | --- | --- | --- | --- |
| | Increasing temperatures and heat waves | Increasing precipitation or heavy downpours | Decreasing water availability | Increasing wildfire | Increasing sea level rise and storm surge | Increasing frequency of intense hurricanes |
| **THERMOELECTRIC POWER GENERATION** | | | | | | |
| Hardening | • Increase or install additional generation capacity<br>• Install additional cooling capacity to existing facilities | • Enhance levees and floodwalls<br>• Install waterproofing measures such as concrete moat walls, floodgates and watertight doors, sluice gates, reinforced walls, pressure resistant/submarine-type doors in deep basements, expansive polymer foam in conduits, submersible pumps<br>• Elevate critical equipment | • Install water-saving cooling technology (e.g., closed-loop cooling, hybrid wet–dry cooling, dry cooling)<br>• Install equipment capable of using alternate water sources (e.g., brackish groundwater, municipal wastewater) for cooling<br>• Install generation technologies with minimal/no water needs (e.g., wind, PV solar) | | • Install sea walls, riprap, and natural barriers such as vegetation<br>• Install waterproofing measures, such as concrete moat walls, floodgates and watertight doors, sluice gates, reinforced walls, pressure-resistant/submarine-type doors in deep basements, expansive polymer foam in conduits, submersible pumps<br>• Elevate critical equipment | • Reinforce elevated structures (e.g., cooling towers, water towers, smokestacks, etc.) for greater wind loading and potential wind-driven debris |
| Planning and operations | • Update integrated resource plans to account for reduced available generation capacity from higher temperatures<br>• See electricity demand section, below | • Update design, siting, and operations plans to account for possibility of increasing floods | • Secure back-up water supply in case of low flow conditions<br>• Install monitoring systems on source water supplies<br>• Develop operating procedures for low water conditions | | • Update design, siting, and operations plans to account for SLR | • Develop alternative fuel delivery options<br>• Maintain larger fuel inventory onsite<br>• Apply extreme wind loading design criteria for critical equipment<br>• Develop or update storm plans to account |

| Climate Threats | | | | | | |
|---|---|---|---|---|---|---|
| Type of Measure | Increasing temperatures and heat waves | Increasing precipitation or heavy downpours | Decreasing water availability | Increasing wildfire | Increasing sea level rise and storm surge | Increasing frequency of intense hurricanes |
| | | | • See electricity demand section, below | | | for higher frequency of intense hurricanes |
| **HYDROELECTRIC POWER GENERATION** | | | | | | |
| Hardening | • Install additional cooling capacity to existing facilities | • Reinforce structures and upgrade equipment to accommodate high flow periods | • Increase storage capacity of reservoirs<br>• Increase turbine efficiency and minimize water leaks at existing dams | | | |
| Planning and operations | • Update integrated resource plans to account for reduced available generation capacity<br>• Incorporate thermal predictive models into reservoir-level forecasts<br>• See electricity demand section, below | • Update design and operation plans to account for altered precipitation patterns (e.g., heavy streamflow events, reduced snowpack, summer drought) | • Develop integrated water management plan that accounts for changing water availability<br>• Manage reservoir capacity (e.g., maintain higher winter carryover storage levels, reduce conveyance flows in canals and flumes, and reduce discretionary reservoir water releases)<br>• Install monitoring systems on rivers with telemetry to increase data availability, trending, and station response times<br>• Develop operating procedures for low water conditions<br>• Improve forecasts of snowmelt timing based | | | |

| Climate Threats | | | | | | |
|---|---|---|---|---|---|---|
| Type of Measure | Increasing temperatures and heat waves | Increasing precipitation or heavy downpours | Decreasing water availability | Increasing wildfire | Increasing sea level rise and storm surge | Increasing frequency of intense hurricanes |
| | | | on snowpack and temperature trends • See electricity demand section, below | | | |
| **BIOENERGY AND RENEWABLE POWER GENERATION** | | | | | | |
| Hardening | • Increase or install additional generating capacity | • Enhance levees and floodwalls • Elevate critical equipment | • Use alternative water supplies at biorefineries (e.g., degraded water or wastewater) • Employ sustainable agriculture methods including crop diversification, crop rotation | | • Install sea walls, riprap, and natural barriers such as vegetation • Elevate critical equipment or enclose equipment in submersible casings | |
| Planning and operations | • Update design plans for increasing temperatures | • Develop alternative fuel delivery options • Maintain larger fuel inventory onsite | • Update plans for securing water, considering decreasing water availability | • Account for increased wildfire risk when siting facilities • Incorporate increased wildfire risk into forest management practices, such as frequency of prescribed burns and reduction of hazardous fuels to prevent uncontrolled fire depleting woody biomass resources | • Update design, siting, and operations plans to account for SLR | • Develop alternative fuel delivery options • Maintain larger fuel inventory onsite • Apply extreme wind loading design criteria • Develop or update storm plans to account for higher frequency of intense hurricanes |
| **ELECTRIC TRANSMISSION AND DISTRIBUTION** | | | | | | |
| Hardening | • Limit customers affected by outages by installing additional substations and breakaway equipment and by sectionalizing fuses; develop island– | • Increase redundancy in transmission system • Enhance levees and floodwalls • Limit customers affected by outages by | | • Increase redundancy in transmission system • Limit customers affected by outages by installing additional substations and | • Install sea walls, riprap, and natural barriers such as vegetation • Limit customers affected by outages by installing additional | • Increase redundancy in transmission system • Limit customers affected by outages by installing technology such as microgrids, |

| Climate Threats | | | | | | |
|---|---|---|---|---|---|---|
| Type of Measure | Increasing temperatures and heat waves | Increasing precipitation or heavy downpours | Decreasing water availability | Increasing wildfire | Increasing sea level rise and storm surge | Increasing frequency of intense hurricanes |
| | • able "microgrids" with distributed generation<br>• Upgrade transformers (e.g., forced-air or forced-oil cooling)<br>• Install smart grid devices that to speed identification of faults and service restoration<br>• Increase or install additional transmission capacity<br>• Install breakable links and towers designed to tolerate lateral movement of foundation in event of uneven permafrost thaw and frost heave<br>• Install additional cooling capacity to existing facilities | • installing technology such as microgrids, additional substations, sectionalizing fuses, and breakaway equipment<br>• Underground critical transmission and distribution lines<br>• Install waterproofing measures, such as floodgates and watertight doors, sluice gates, reinforced walls, pressure-resistant/ submarine-type doors in deep basements, expansive polymer foam in conduits<br>• Elevate or relocate critical equipment | | • breakaway equipment and by sectionalizing fuses; develop island-able "microgrids" with distributed generation<br>• Replace wood poles and support structures with fire-resistant materials (e.g., steel or concrete)<br>• Install smart grid devices to speed identification of faults and service restoration | • substations and breakaway equipment and by sectionalizing fuses; develop island-able "microgrids" with distributed generation<br>• Replace wood poles and support structures with stronger materials (e.g., steel or concrete)<br>• Elevate or relocate critical equipment<br>• Install smart grid devices to speed identification of faults and service restoration | • additional substations, sectionalizing fuses, and breakaway equipment<br>• Replace wood poles and support structures with stronger materials (e.g., steel or concrete)<br>• Underground critical transmission and distribution lines<br>• Replace ceramic insulators with polymer<br>• Install smart grid devices to speed identification of faults and service restoration<br>• Utilize mobile transformers and substations |
| Planning and operations | • Develop best operating practices for equipment at high temperatures<br>• Include extreme temperature scenarios in future grid planning<br>• Deploy future equipment and lines with higher design temperatures | • Site equipment in areas less prone to flooding<br>• Install water-level monitoring systems and communications equipment inside vulnerable substations | | • Site equipment in areas less prone to wildfire<br>• Enhance vegetation management (e.g., tree trimming, forest thinning, and prescribed burning)<br>• Develop fire response plans and tools; coordinate with local partners | • Site equipment in areas less prone to coastal flooding<br>• Install water-level monitoring systems and communications equipment inside vulnerable substations<br>• Update siting and operations plans to account for SLR | • Apply extreme wind loading design criteria to critical infrastructure<br>• Site equipment further from coast<br>• Enhance vegetation management<br>• Update storm plans to account for higher |

| Type of Measure | Increasing temperatures and heat waves | Increasing precipitation or heavy downpours | Decreasing water availability | Increasing wildfire | Increasing sea level rise and storm surge | Increasing frequency of intense hurricanes |
|---|---|---|---|---|---|---|
| **Climate Threats** | | | | | | |
| | | | | • Develop firefighting compounds safe to use near active power lines | | frequency of intense hurricanes |
| **ELECTRICITY DEMAND** | | | | | | |
| Hardening | • Implement weatherization programs<br>• Install energy efficient equipment<br>• Increase generation and transmission capacity<br>• Invest in grid-scale energy storage systems | | • Implement water and energy efficient technologies and practices to reduce energy demand for water production, pumping, and filtration | | | |
| Planning and operations | • Update resource plans to accommodate projected increases in CDDs and decreases in HDDs<br>• Implement programs that incentivize and encourage energy efficiency<br>• Implement load management and demand side response programs | | • Emphasize water efficiency in buildings, industrial processes, municipal utilities, and in other areas to reduce energy demand for water production, pumping, and filtration | | | |
| **SUPPLY CHAIN: FUEL TRANSPORT** | | | | | | |
| Hardening | • Engineer structures in permafrost areas with design criteria suited for warming<br>• Insulate or ventilate underlying permafrost, | • Enhance levees and floodwalls<br>• Elevate critical equipment | • Use alternative water supplies, such as degraded water, wastewater, brackish water, or produced water | • Install emergency backup power, such as diesel generators, for critical operations | • Install sea walls, riprap, and natural barriers such as vegetation<br>• Elevate critical equipment | • Install emergency backup generators for critical operations<br>• Incorporate more robust design specifications for |

| Type of Measure | Climate Threats | | | | | |
|---|---|---|---|---|---|---|
| | Increasing temperatures and heat waves | Increasing precipitation or heavy downpours | Decreasing water availability | Increasing wildfire | Increasing sea level rise and storm surge | Increasing frequency of intense hurricanes |
| | such as construction of a gravel pad of appropriate depth or the use of thermal piles | • Install emergency backup generators for critical operations | | | | equipment in hurricane zones<br>• Locate rigs on more stable areas of sea floor<br>• Brace vulnerable equipment to protect from wind damage |
| Planning and operations | • Update design and operations guides for equipment operating in Arctic Alaska | • Update design, siting, and operations plans to account for heavy runoff and possible increasing floods | • Update plans for securing water to consider decreasing water availability | • Update wildfire response plans to account for increasing frequency and severity | • Update siting and operations plans to account for SLR | • Update design criteria for new equipment in hurricane zones to account for extreme wind loading<br>• Update engineering and operations guidance and storm plans to account for higher frequency of intense hurricanes |

99

# APPENDIX B REFERENCES

[1] DOE. 2015. *Climate Change and the U.S. Energy Sector: Regional Vulnerabilities and Resilience Solutions.* Washington, DC: DOE. October. http://energy.gov/sites/prod/files/2015/10/f27/Regional_Climate_Vulnerabilities_and_Resilience_Solutions_0.pdf.